Processing of Advanced Ceramics

Yujun Zhang

Beijing
Metallurgical Industry Press
2014

Abstract

This book presents a brief introduction to the processing of advanced ceramics. There are four chapters in this book: Advanced Ceramics and their Applications, Powder Processing, Shape-forming, and Sintering. It introduces some new methods in ceramic powder synthesis, shape-forming and sintering of ceramics. This book also especially provides some processing detail related to the production of alumina, silicon carbide, silicon nitride and boron carbide ceramics with combination of author's experience.

This book can be used as a text book and a reference book for the students who are studying the inorganic non-metal materials science and engineering, and for engineers and researchers who are working in the field of advanced ceramics.

图书在版编目(CIP)数据

先进陶瓷工艺 = Processing of advanced ceramics：英文/张玉军编著. —北京：冶金工业出版社，2014.10
ISBN 978-7-5024-6736-4

Ⅰ.①先⋯ Ⅱ.①张⋯ Ⅲ.①陶瓷—工艺学—英文 Ⅳ.①TQ174.6

中国版本图书馆 CIP 数据核字(2014)第 231777 号

出 版 人　谭学余
地　　址　北京市东城区嵩祝院北巷39号　邮编 100009　电话 (010)64027926
网　　址　www.cnmip.com.cn　电子信箱 yjcbs@cnmip.com.cn
责任编辑　于昕蕾　美术编辑　吕欣童　版式设计　孙跃红
责任校对　禹　蕊　责任印制　李玉山

ISBN 978-7-5024-6736-4
冶金工业出版社出版发行；各地新华书店经销；北京佳诚信缘彩印有限公司印刷
2014年10月第1版，2014年10月第1次印刷
169mm×239mm；13.75 印张；268 千字；212 页
52.00 元

冶金工业出版社　投稿电话　(010)64027932　投稿信箱 tougao@cnmip.com.cn
冶金工业出版社营销中心　电话 (010)64044283　传真 (010)64027893
冶金书店　地址　北京市东四西大街46号(100010)　电话 (010)65289081(兼传真)
冶金工业出版社天猫旗舰店　yjgy.tmall.com

(本书如有印装质量问题,本社营销中心负责退换)

Preface

The objective of this book is to offer some background knowledge for the students who are majoring in the inorganic non-metallic materials science and engineering in the Chinese domestic universities. The author reads some original English books concerning to ceramic science and engineering, and tries to choice one, but find no one is very suitable, because the content of every book is too abundant.

It is very difficult to write an English professional textbook. The author has redacted this book by excerpting the contents from *Ceramic Innovations in the 20th Century* (John B. Wachtman Jr., published by The American Ceramic Society, 1999), *Advanced Technical Ceramics* (Shigeyuki Somiya, published by Academic Press Japan Inc., 1984), *High-tech Ceramics* (Gernot Kostorz, published by Academic Press Inc., 1989), *Dictionary of Ceramics* (Third edition, Arthur Dodd, published by the Institute of Materials, 1994), *Ceramic Processing and Sintering* (Second edition, M. N. RAHAMAN, published by Marcel Dekken Inc., 2003), *Ceramic Technology and Processing* (Alan G. King, published by Noyes Publications, 2002), *Ceramic Materials, Science and Engineering* (C. Barry Carter and M. Grant Norton, published by Springer Science + Business media, LLC., 2007), *Modern Ceramic Engineering* (Third Edition, David W. Richerson, published by Taylor & Francis Group, 2006), *Sintering, Densification, Grain Growth, and Microstructure* (Suk-Joong L. Kang, published by Elsevier Butterworth-Heinemann, 2005), and references recently published. Also, we have specially introduced the technology and processing of alumina ceramics, silicon nitride ceramics and silicon carbide ceramics production in this book.

This book presents a brief introduction to the advanced ceramic processing and includes four chapters: Advanced Ceramics and Their Applications, Powder Processing, Shape-Forming, and Sintering.

It is the author's wish that "Processing of Advanced Ceramics" will promote the understanding of the technology and processing of advanced ceramics for the students who are studying the Inorganic Non-metal Materials Science and Engineering, and for engineers and researchers who are working in the field of advanced ceramics.

The author would like to thank to Prof. Hongyu Gong for offering help of tables and data in Chapter 4.

Dr. Prof. Yujun Zhang
School of Materials Science and Engineering
Shandong University
August 2014

Contents

1 Advanced Ceramics and Their Applications 1

1.1 Definition 1
1.2 Ceramics Classification 2
1.3 Progress of Advanced Ceramics 4
1.4 Basic Properties of Advanced Ceramics 6
    1.4.1 Oxide Ceramics 6
    1.4.2 Non-Oxide Ceramics 6
1.5 Applications of Advanced Ceramics 22
Study Guide 25
References 25

2 Powder Processing 27

2.1 Raw Materials 27
    2.1.1 Oxides 27
    2.1.2 Nonoxides 33
    2.1.3 Raw Material Selection 42
2.2 Preparation of Powders 45
    2.2.1 Mechanical Methods to Prepare Powders 47
    2.2.2 Powder Synthesis by Chemical Methods 61
    2.2.3 Powders by Vapor-phase Reactions 73
    2.2.4 Spray Drying 77
2.3 Characterizing Powders 81
    2.3.1 Characterizing Powders by Microscropy 81
    2.3.2 Characterizing Powders by Light Scattering 82
    2.3.3 Characterizing Powders by X-Ray Diffraction 82
    2.3.4 Characterizing Powders by Surface Area (The BET Method) 84
    2.3.5 Determining Particle Composition and Purity 84

Study Guide ········ 86
References ········ 86

3 Shape-Forming ········ 89

3.1 Additives ········ 90
3.1.1 Solvents ········ 90
3.1.2 Dispersants ········ 92
3.1.3 Binder ········ 95
3.1.4 Plasticizers ········ 97
3.1.5 Other Additives ········ 97
3.2 Pressing ········ 98
3.2.1 Uniaxial Pressing ········ 98
3.2.2 Isostatic Pressing ········ 101
3.3 Casting ········ 106
3.3.1 Slip Casting ········ 106
3.3.2 Tape Casting ········ 111
3.3.3 Gel Casting ········ 114
3.3.4 Electrophoretic deposition ········ 118
3.4 Plastic Forming ········ 120
3.4.1 Extrusion ········ 121
3.4.2 Injection ········ 125
3.4.3 Roll Forming ········ 130
Study Guide ········ 133
References ········ 134

4 Sintering ········ 136

4.1 Kiln and Furnace ········ 136
4.1.1 Classification of Ceramic Furnaces (Kilns) ········ 137
4.1.2 Batch Kilns ········ 139
4.1.3 Continuous Kiln ········ 145
4.1.4 Heat Elements ········ 147
4.1.5 Insulation and Refractories ········ 150
4.1.6 Furniture and Crucibles ········ 151

4.1.7	Measuring Temperature	152
4.2	**Sintering of Advanced Ceramics**	157
4.2.1	Solid-state Sintering and Liquid-sintering	159
4.2.2	Pressureless Sintering	166
4.2.3	Hot Pressing Sintering	175
4.2.4	Hot Isostatic Pressing	179
4.2.5	Gas Pressing Sintering	180
4.2.6	Reaction Sintering	182
4.2.7	Spark Plasma Sintering	185
4.2.8	Microwave Sintering	191
4.2.9	Compare of Several New Sintering Methods	199
4.3	**Control of Sintering**	201
4.3.1	Firing Curve	201
4.3.2	Atmosphere	202
4.3.3	Sintering Problems	203
Study Guide		205
References		205

1 Advanced Ceramics and Their Applications

Materials science and engineering plays a vital role in this modern age of science and technology. Various kinds of materials are used in industry, housing, agriculture, transportation, etc to meet the plant and individual requirements.

In classical materials science, materials are generally grouped into four categories: metals, polymers, ceramics, and composites. The first three are based primarily on the nature of the interatomic bonding, the fourth on the materials structure.

1.1 Definition

Ceramics are usually associated with "mixed" bonding—a combination of ionic and/or covalent, and sometimes metallic. Many (but not all) are crystalline in nature and are compounds formed between metallic and nonmetallic elements such as aluminum and oxygen (alumina-Al_2O_3), calcium and oxygen (calcia-CaO), and silicon and nitrogen (silicon nitride-Si_3N_4). However, diamond and graphite are also classified as ceramics.

The word ceramic was initially derived from the Greek term keramos, meaning "a potter" or "pottery." Keramos in turn is related to an older Sanskrit root meaning "to burn." Thus the early Greeks used the term to mean "burned stuff" or "burned earth" when referring to products obtained through the action of fire upon earthy materials.

In the past, this word "Ceramic" was applied to dishes, made of clay, and to be cooked afterwards; but nowadays, this word comprises all products made of fictile clay and bat, such as brick, tile, clay pipe and all fire-proofs made of bat, vitreous china and various species of porcelain, as well as china-ware.

By the latter of the 20th century, scientists and engineers had learned to synthesize many new ceramics, sometimes by chemical methods that did not quite fit the older definitions. W. D. Kingery suggested in his classic text "Introduction to Ceramics" a new definition: "**A ceramic is a nonmetallic, inorganic solid.**" This definition essentially says that a ceramic is anything that is not an organic material or a metal[1].

For the word ceramic, different literatures give different definitions. Ceramic is a hard brittle material made by firing clay and similar substances (*World English Diction-*

ary). Ceramic is any of various hard, brittle, heat-resistant and corrosion-resistant materials made by shaping and then firing a nonmetallic mineral, such as clay, at a high temperature (technical ceramics, advanced toughened ceramics, and precise ceramic parts). Ceramic is any of various hard, brittle, heat-and corrosion-resistant materials made typically of metallic elements combined with oxygen or with carbon, nitrogen, or sulfur. Most ceramics are crystalline and are poor conductors of electricity (*Science Dictionary*).

What are ceramics? The American Ceramic Society, the National Institute of Ceramic Engineers, the Ceramic Educational Council, the American Society for Testing and Materials, and other learned societies and associations define ceramics as inorganic, non-metallic materials that are subjected to elevated temperatures during their processing and manufacture-temperatures of 540℃ and above (*Dictionary of Ceramic Science and Engineering*, Loran S. Q'Bannon, Plenum Press).

Encyclopedia gives a definition that ceramics are broadly defined as inorganic, non-metallic materials that exhibit such useful properties as high strength and hardness, high melting temperatures, chemical inertness, and low thermal and electrical conductivity but that also display brittleness and sensitivity to flaws.

It is also not possible to define ceramics, or indeed any class of material, in terms of specific properties. We cannot say "ceramics are brittle" because some can be superplastically deformed. We cannot say "ceramics are insulators" because some have high electrical conductivity. We cannot say "ceramics are poor conductors of heat" because diamond has the highest thermal conductivity of any known material.

Now it is difficult to give a precise definition for ceramic, however, it would be regarded to the most widely accepted definition that ceramics is any of a class of inorganic, nonmetallic products which are subjected to an high temperature during manufacture or use and which are produced with natural mineral materials and/or synthetic materials and/or chemical products as raw materials.

1.2 Ceramics Classification

The subject of ceramics covers a wide range of materials. Recent attempts have been made to divide it into two parts: **traditional silicate products** and **advanced ceramics**[2].

The term Advanced Ceramics is opposite in meaning to "Traditional" or "Classical" ceramics. In the past, Advanced Ceramics were often confused with New or Newer Ceramics, Modern Ceramics, Special Ceramics and so on. Furthermore, Fine Ceramics, at least in the USA and Europe, is synonymous with Fine Grain Ceramic Products and/or Fine Grain Porcelain; Fine Ceramics in Japan is similar to what we understand as Advanced Ceramics[3]. In China, the terms advanced ceramics and high-technology ceramics are often used.

Traditional silicate products, made from common, naturally occurring minerals such as clay and sand, consist of structural clay products (i. e. brick, roofing tile, clay floor and wall tile, flue linings), traditional ceramics (i. e. whiteware), refractory (brick and monolithic products are used in iron and steel, non-ferrous metals, glass, cements, ceramics, energy conversion, petroleum, and chemicals industries), glass (windows glass, container glass, pressed and blown glass, glass fibers, and advanced/special glass fibers), abrasive (garnet, diamond, silicon carbide, fused alumina, etc are used for grinding, cutting, polishing, or pressure blasting of materials), cements (used to produce concrete roads, bridges, buildings, dams), enamels etc.

Traditional ceramics can be classified as building ceramics (glazed tiles, floor tiles), sanitary-wares, decorative ceramics, electrical porcelain, industrial ceramics, dinnerware etc.

Advanced ceramics are defined as those ceramics that exhibit a high degree of industrial efficiency through their carefully controlled composition and designed microstructures, used in technical applications where their mechanical, thermal, electrical and/or optical properties are important. Containing a variety of ingredients and manipulated by a variety of processing techniques, advanced ceramics are made into a wide range of industrial products, from common ceramic water filter to nuclear fuel pellets.

There are many possible approaches to classifying ceramics. They can be grouped, for instance, by their chemical composition, their mineral content, the processing methods used in their production, their properties, or their uses. F. Thümmler[4] has classified the advanced ceramics by their main function, as Table 1.1. Nevertheless, advanced ceramics can be also classified as engineering ceramics and functional ceramics by their properties.

Table 1.1 Classification of advanced ceramics[4]

Main function	Properties required	Applications (examples)
Thermal	High-temperature and thermal shock resistance, thermal conductivity (high or low, respectively)	High-temperature components, burner nozzles, heat exchangers, heating elements, non-iron metallurgy, insulating parts, thermal barrier coatings
Mechanical	Long-term, high-temperature resistance, fatigue, thermal shock, wear resistance	Wear parts, sealings, bearings, cutting tools, engine, motor and gas turbine parts, thermal barrier coatings

Continues Table 1.1

Main function	Properties required	Applications (examples)
Chemical, biological	Corrosion resistance, bio-compatibility	Corrosion protection, catalyst carriers, environmental protection, sensors, implants (joints, teeth, etc)
Electrical, magnetic	Electrical conductivity (high or low, respectively), semi-conducting, piezo-, thermoelectricity, dielectrical properties	Heating elements, insulators, magnets, sensors, IC-packages, substrates, solid electrolytes, piezo-electrics, dielectrics, super conductors
Optical	Low absorption coefficient	Lamps, windows, fibre optics, IR-optics
Nuclear	Irradiation resistance, high absorption coefficient, high-temperature resistance, corrosion resistance	Fuel and breeding elements, absorbers, shields, waste conditioning

Engineering ceramics, also called as structural ceramics, is a subject of technical ceramics, including principally ceramics with superior mechanical properties, such as great strength, abrasion resistance, high level of elasticity, enhanced hardness, heat resistance, and lubricating ability. These ceramics have found applications in ceramic engines, cutting tools, grinding materials, and materials for bearings.

Functional ceramics is a subject of technical ceramics, including principally ceramics with superior thermal, electrical, magnetic, and optical properties. Functional ceramics includes bio-ceramics, electronic ceramics, magnetic ceramics, optical ceramics, nuclear and environmental ceramics, superconducting ceramics, electro-optic ceramics etc.

1.3 Progress of Advanced Ceramics

"Ceramics" are the materials which are often misunderstood as materials used merely for pottery and decorative objects. Even though the word ceramic is derived from the Greek word "Keramos", what many people don't realize is that ceramics play an important role almost everywhere you see and many times in places that you can't. Besides the everyday objects of glassware and floor tiles, the ceramics of today are critical in helping computers and other electronic devices operate, in medical devices for improving people's health in various ways, in providing global telecommunications, and in protecting soldiers and vehicles during combat.[5]

Various "advanced ceramic" products are manufactured by combining high-purity chemicals into desired shapes and then heating them to very high temperature. The shaped ceramic products thus made can have many desirable properties such as heat resistance, hardness, strength, low electrical conductivity, and unique electro-mechanical characteristics. Thus advanced ceramics are ceramics which are made by tightly controlled methods and therefore they exemplify an "advancement" over the general definition. As a consequence of these refined methods, a new class of ceramics called "**advanced ceramics**" is born.

One of the first uses of advanced ceramics was for corrosion-resistant stoneware vessels in the chemical industry as early as the 1750s. Then came porcelain, which was first used in dentistry in the 1850s. With the invention of electric light in the 19th century, ceramic materials based on porcelain for electrical insulation were developed. This was followed by the blooming of the radio and television broadcasting industry in the 20th century, which needed special heat resistant materials that could withstand the high-frequency electromagnetic fields. As a result, electro-ceramics such as steatite were developed. Subsequently, other electro-ceramics such as magnetic ceramics (ferrites) were developed, followed by capacitor ceramics (titanates) and electro-mechanical ceramics (piezoelectric ceramics). In the later part of the 20th century, the need for protecting tiny transistors and ICs (integrated circuits) from ambient conditions led to the development of ceramic packaging materials which facilitated further miniaturization.

Concurrent with the development of electro-ceramics, another sub-class of advanced ceramics which came to be called structural ceramics progressed, which had high structural and chemical integrity characterized by properties such as extremely high hardness, stiffness, and heat and chemical resistance. These structural ceramics found applications in various industries, for example in the space industry as heat and wear resistant tiles and nose cones on space shuttles, in the aerospace industry as bearings and turbine rotors, in the chemical industry as chemical resistant seals and conduits, in the defence industry as bullet-proof vests and armor plates for vehicles, in the biomedical industry as hip-joints, knee-joints and orbital implants, and so on.

In the past three decades, breakthroughs in advanced ceramics have enabled significant new technology capabilities that are now having far-reaching impacts on the world economy and society. For example, ceramic catalytic converters are responsible for reducing automobile emissions and long-life bearings are used in a wide range of applications to improve performance and reduce friction.

Advanced ceramics continue to be developed even further; new ceramics and their combinations are constantly being improved and newer products are continually being introduced in various industries.

The progress in advanced ceramics is so rapid that the advanced ceramics of today are strikingly different from those made even a few years ago. Ceramic engineers eagerly anticipate further rapid developments of newer ceramic materials and their combinations that will find even more exciting applications in the future.

1.4 Basic Properties of Advanced Ceramics[1]

As mentioned above, advanced ceramics can be classified by chemical composition, properties, or applications. In the present section we list the basic properties of some advanced ceramic materials by their classification of chemical composition.

1.4.1 Oxide Ceramics

Oxides are the most common constituents of all ceramics, traditional and technical. Table 1.2 presents a systematic listing of their composition, structure and properties.

1.4.2 Non-Oxide Ceramics

The non-oxide ceramics comprise essentially borides, carbides, nitrides, and silicides. Like oxide ceramics they have two kinds of uses, which frequently overlap: application of their physical properties and of their refractory high-temperature properties.

The two dominating design variables to be considered for high-temperature materials are hardness and thermal conductivity as a function of temperature. Figure 1.1 gives a survey. It is obvious that borides and carbides are superior to most oxide ceramics. Depending upon the microstructure, SiC has a higher hardness than that of β-Si_3N_4, but it decreases somewhat more rapidly, with increasing temperature. Both Si_3N_4 and SiC ceramics possess a high thermal conductivity and thus excellent thermal-shock resistance.

1.4.2.1 Borides
Some properties of borides and boride-based ceramics are listed in Table 1.3.

1.4.2.2 Carbides
Some properties of carbides and carbide-based high-temperature ceramics are listed in Table 1.4 and Table 1.5. It should be noted that carbides also play a major role as

---

[1] In this section, Table 1.2 ~ Table 1.9 selected from *Springer Handbook of Condensed Matter and Materials Data* (Springer Berlin Heidelberg, 2005) edited by Werner Martienssen and Hans Warlimont, 431 ~ 476, with some cut.

1.4 Basic Properties of Advanced Ceramics

Table 1.2 Physical properties of oxides and oxide-based high-temperature ceramics

IUPAC name (synonyms name)	Theoretical Chemical formula, Relative molecular mass	Density ρ/kg·m^{-3}	Electrical resistivity ρ/$\mu\Omega$·cm	Melting point /°C	Thermal conductivity κ/W·(m·K)$^{-1}$	Specific heat capacity c_p /J·(kg·K)$^{-1}$	Coefficient Linear thermal expansion α/K^{-1}
Aluminium sesquioxide (alumina)	α-Al$_2$O$_3$, 101.961	3987	2×10^{23}	2054	35.6~39	795.5~880	7.1×10^{-6} ~ 8.3×10^{-6}
Beryllium monoxide (beryllia)	BeO, 25.011	3008~3030	10^{22}	2550~2565	245~250	996.5	7.5×10^{-6} ~ 9.7×10^{-6}
Calcium monoxide (calcia, lime)	CaO, 56.077	3320	10^{14}	2927	8~16	753.1	3.88×10^{-6}
Cerium dioxide (ceria, cerianite)	CeO$_2$, 172.114	7650	10^{10}	2340	—	389	10.6×10^{-6}
Chromium Oxide (eskolaite)	Cr$_2$O$_3$, 151.990	5220	1.3×10^9 (346°C)	2330	—	921.1	10.9×10^{-6}
Dysprosium oxide (dysprosia)	Dy$_2$O$_3$, 373.00	8300	—	2408	—	—	7.74×10^{-6}
Europium oxide (europia)	Eu$_2$O$_3$, 351.928	7422	—	2350	—	—	7.02×10^{-6}
Hafnium dioxide (hafnia)	HfO$_2$, 210.489	9680	5×10^{15}	2900	1.14	121	5.85×10^{-6}
Gadolinium oxide (gadolinia)	Gd$_2$O$_3$, 362.50	7630	—	2420	—	276	10.44×10^{-6}
Lanthanum oxide (lanthana)	La$_2$O$_3$, 325.809	6510	10^{14} (550°C)	2315	—	288.89	11.9×10^{-6}
Magnesium monoxide (magnesia)	MgO, 40.304	3581	1.3×10^{15}	2852	50~75	962.3	11.52×10^{-6}
Samarium oxide (samaria)	Sm$_2$O$_3$, 348.72	7620	—	2350	2.07	331	10.3×10^{-6}
Silicon dioxide (silica, α-quartz)	α-SiO$_2$, 60.085	2202~2650	10^{20}	1710	1.38	787	0.55×10^{-6}
Tantalum pentoxide (tantalite)	Ta$_2$O$_5$, 441.893	8200	10^{12}	1882	—	301.5	—
Thorium dioxide (thoria, thorianite)	ThO$_2$, 264.037	9860	4×10^{19}	3390	14.19	272.14	9.54×10^{-6}
Titanium dioxide (rutile, titania)	TiO$_2$, 79.866	4240	10^{19}	1855	10.4 ($\parallel c$), 7.4 ($\perp c$)	711	7.14×10^{-6}
Uranium dioxide (uraninite)	UO$_2$, 270.028	10960	3.8×10^{10}	2880	10.04	234.31	11.2×10^{-6}

Continues Table 1.2

IUPAC name (synonyms name)	Theoretical Chemical formula, Relative molecular mass	Vickers hardness H_V (Mohs H_M)	Compressive strength σ/MPa	Flexural strength τ/MPa	Young's modulus E/GPa	Density ρ/kg·m^{-3}	Electrical resistivity ρ/μΩ·cm	Melting point /°C	Thermal conductivity κ/W·(m·K)$^{-1}$	Specific heat capacity c_p /J·(kg·K)$^{-1}$	Coefficient Linear thermal expansion α/K^{-1}	Other physicochemical properties, corrosion resistance, and uses
Yttrium oxide (yttria)	Y_2O_3, 225.81	2100~3000 (H_M9)	2549~3103	282	365~393	5030	—	2439	—	439.62	8.10×10^{-6}	White and translucent; hard material used as abrasive for grinding. Excellent electrical insulator and also wear-resistent. Insoluble in water, and in strong mineral acids, readily soluble in strong solutions of alkali metal hydroxides, attacked by HF and NH_4HF_2. Owing to its corrosion resistance under an inert atmosphere in molten metals such as Mg, Ca, Sr, Ba, Mn, Sn, Pb, Ga, Bi, As, Sb, Hg, Mo, W, Co, Ni, Pd, Pt, and U, it is used for crucibles for these liquid metals. Alumina is readily attacked under an inert atmosphere by molten metals such as Li, Na, Be, Al, Si, Ti, Zr, Nb, Ta, and Cu. Maximum service temperature 1950℃
Zirconium dioxide (baddeleyite)	ZrO_2, 123.223	1500 (H_M9)	1551	241~250	296.5~345	5850	—	2710	—	711	7.56×10^{-6}	Forms white or grayish ceramics. It is readily absorbs CO_2 and water from air to form calcium carbonate and slaked lime. It reacts readily with water to give $Ca(OH)_2$. Volumetric expansion co-efficient 0.225×10^{-9} K^{-1}. It exhibits outstanding corrosion resistance in the following liquid metals: Li and Na
		560 (H_M4.5)	—	—	—							Pale yellow cubic crystals. Abrasive for polishing glass; used in interference filters and antireflection coatings. Insoluble in water, soluble in H_2SO_4 and HNO_3, but insoluble in dilute acids
		(H_M6)	589	—	181							
		(H_M>8)	—	—	—							
		—	—	—	—							
		780~1050	—	—	57							
		480	—	—	124							Insoluble in water, soluble in dilute strong mineral acids
		—	—	—	—							

1.4 Basic Properties of Advanced Ceramics

Continues Table 1.2

Young's modulus E/GPa	Flexural strength τ/MPa	Compressive strength σ/MPa	Vickers hardness H_V (Mohs H_M)	Other physicochemical properties, corrosion resistance, and uses
303.4	441	1300~1379	750 (H_M5.5~6)	Forms ceramics with a high reflection coefficient in the visible and near-UV region. Used in linings for steelmaking furnaces and in crucibles for fluoride melts. Very slowly soluble in pure water; but soluble in dilute strong mineral acids. It exhibits outstanding corrosion resistance in the following liquid metals: Mg, Li, and Na. It is readily attacked by molten metals such as Be, Si, Ti, Zr, Nb, and Ta. MgO reacts with water, CO_2, and dilute acids. Maximum service temperature 2400℃. Transmittance of 80% and refractive index of 1.75 in the IR region from 7 to 300μm
183	—	—	438	Colorless amorphous (fused silica) or crystalline (quartz) material having a low thermal expansion coefficient and excellent optical transmittance in the far UV. Silica is insoluble in strong mineral acids and alkalis except HF, concentrated H_3PO_4, NH_4HF_2 and concentrated alkali metal hydroxides. Owing to its good corrosion resistance to liquid metals such as Si, Ge, Sn, Pb, Ga, In, Tl, Rb, Bi, and Cd, it is used in crucibles for melting these metals. Silica is readily attacked under an inert atmosphere by molten metals such as Li, Na, K, Mg, and Al. Quartz crystals are piezoelectric and pyroelectric. Maximum service temperature 1090℃
72.95	310	680~1380	550~1000 (H_M7)	Dielectric used in film supercapacitors. Tantalum oxide is a high-refractive index, low-absorption material usable for making optical coatings from the near-UV (350 nm) to the IR (8μm). Insoluble in most chemicals except HF, HF-HNO_3 mixtures, oleum, fused alkali metal hydroxides and molten pyrosulfates
—	—	—	—	Corrosion-resistant container material for the following molten metals; Be, Si, Ti, Zr, Nb, Bi. Radioactive
144.8	—	1475	945 (H_M6.5)	White, translucent, hard ceramic material. Readily soluble in HF and in concentrated H_2SO_4, and reacts rapidly with molten alkali hydroxides and fused alkali carbonates. Owing to its good corrosion resistance to liquid metals such as Ni and Mo, it is used in crucibles for melting these metals. Titania is readily attacked under an inert atmosphere by molten metals such as Be, Si, Ti, Zr, Nb, and Ta
248~282	340	800~940	(H_M7~7.5)	Used in nuclear power reactors in sintered nuclear-fuel elements containing either natural or enriched uranium
145	—	—	600(H_M6~7)	Yttria is a medium-refractive-index, low-absorption material usable for optical coating, in the near-UV (300nm) to the IR (12μm) regions. Hence used to protect Al and Ag mirrors. Used for Crucibles containing molten lithium
114.5	—	393	700	Zirconia is highly corrosion-resistant to molten metals such as Bi, Hf, Ir, Pt, Fe, Ni, Mo, Pu, and V, while is strongly attacked by the following liquid metals: Be, Li, Na, K, Si, Ti, Zr, and Nb. Insoluble in water, but slowly soluble in HCl and HNO_3; soluble in boiling concentrated H_2SO_4 and alkali hydroxides and readily attacked by HF. Monoclinic (baddeleyite) below 1100℃, tetragonal between 1100 and 2300℃, cubic (fluorite type) above 2300℃. Maximum service temperature 2400℃
241	—	2068	(H_M6.5) 1200	

Table 1.3 Physical properties of borides and boride-based high-temperature ceramics

IUPAC name	Theoretical chemical formula, relative molecular mass	Density $\rho/\text{kg}\cdot\text{m}^{-3}$	Electrical resistivity $\rho/\mu\Omega\cdot\text{cm}$	Melting point /℃	Thermal conductivity $\kappa/\text{W}\cdot(\text{m}\cdot\text{K})^{-1}$	Specific heat capacity c_p /J·(kg·K)$^{-1}$	Coefficient of linear thermal expansion α/K^{-1}
Aluminium diboride	AlB_2, 48.604	3190	—	1654	—	897.87	—
Beryllium diboride	BeB_2, 30.634	2420	10000	1970	—	—	—
Chromium diboride	CrB_2, 73.618	5160~5200	21	1850~2100	20~32	712	6.2×10^{-6} ~ 7.5×10^{-6}
Hafnium diboride	HfB_2, 200.112	11190	8.8~11	3250~3380	51.6	247.11	6.3×10^{-6} ~ 7.6×10^{-6}
Lanthanum hexaboride	LaB_6, 203.772	4760	17.4	2715	47.7	—	6.4×10^{-6}
Molybdenum diboride	MoB_2, 117.59	7780	45	2100	—	527	7.7×10^{-6}
Silicon hexaboride	SiB_6	2430	200000	1950	—	—	—
Tantalum diboride	TaB_2, 202.570	12540	33	3037~3200	10.9~16.0	237.55	8.2×10^{-6} ~ 8.8×10^{-6}
Titanium diboride	TiB_2, 69.489	4520	16~28.4	2980~3225	64.4	637.22	7.6×10^{-6} ~ 8.6×10^{-6}
Tungsten monoboride	WB, 194.651	15200, 16000	4.1	2660	—	—	6.9×10^{-6}
Uranium diboride	UB_2, 259.651	12710	—	2385	51.9	—	9×10^{-6}
Uranium tetraboride	UB_4, 281.273	5350	—	2495	4.0	—	7.0×10^{-6}
Vanadium diboride	VB_2, 72.564	5070	23	2450~2747	42.3	647.43	7.6×10^{-6} ~ 8.3×10^{-6}
Zirconium diboride	ZrB_2, 112.846	6085	9.2	3060~3245	57.9	392.54	5.5×10^{-6} ~ 8.3×10^{-6}

Continues Table 1.3

Young's modulus E/GPa	Flexural strength τ/MPa	Compressive strength σ/MPa	Vickers hardness H_V ($Mohs\ H_M$)	Other physicochemical properties, corrosion resistance, and uses
—	—	—	2500	Phase transition to AlB_{12} at 920℃. Soluble in dilute HCl. Nuclear shielding material
—	—	—	—	Strongly corroded by molten metals such as Mg, Al, Na, Si, V, Cr, Mn, Fe, and Ni. It is corrosion-resistant to the following liquid metals: Cu, Zn, Sn, Rb, and Bi
211	607	1300	1800	
500	350	—	2900	Gray crystals, attacked by HF, otherwise highly resistant
479	126	—	1280	Wear-resistant, semiconducting, thermoionic-conductor films
—	—	—	($H_M > 8$)	Gray metallic powder. Severe oxidation in air above 800℃. Corroded by the following molten metals: Nb, Mo, Ta, and Re
257	—	—		
372~551	240	669	3370 ($H_M > 9$)	Gray crystals, superconducting at 1.26 K. High-temperature electrical conductor, used in the form of a cermet as a crucible material for handling molten metals such as Al, Zn, Cd, Bi, Sn, and Rb. It is strongly corroded by liquid metals such as Tl, Zr, V, Nb, Ta, Cr, Mn, Fe, Co, Ni, and Cu. Begins to be oxidized in air above 1100~1400℃ corrosion-resistant in hot concentrated brines. Maximum operating temperature 1000℃ (reducing environment) and 800℃ (oxidizing environment)
—	—	—	($H_M\ 9$)	Black powder
—	—	—	1390	
268	—	—	($H_M\ 8~9$)	Wear-resistant, semiconducting films
343~506	305	—	1900~3400 ($H_M\ 8$)	Gray metallic crystals, excellent thermal-shock resistance, greatest oxidation inertness of all refractory hard metals. Hot-pressed material is used in crucibles for handling molten metals such as Zn, Mg, Fe, Cu, Zn, Cd, Sn, Pb, Rb, Bi, Cr, brass, carbon steel, and cast iron, and also molten cryolite, yttria, zirconia, and alumina. It is readily corroded by liquid metals such as Si, Cr, Mn, Co, Ni, Nb, Mo, and Ta, and attacked by molten salts such as Na_2O, alkali carbonates, and NaOH. Severe oxidation in air occurs above 1100~1400℃. Stable above 2000℃ under inert or reducing atmosphere

Table 1.4 Physical properties of carbides and carbide-based high-temperature ceramics

IUPAC name (synonyms name)	Theoretical chemical formula, relative molecular mass	Density ρ/kg·m^{-3}	Electrical resistivity $\rho/\mu\Omega\cdot$cm	Melting point /°C	Thermal conductivity κ/W·(m·K)$^{-1}$	Specific heat capacity c_p /J·(kg·K)$^{-1}$	Coefficient of linear thermal expansion α/K^{-1}
Aluminium carbides	Al$_4$C$_3$, 143.959	2360	—	2798	n.a	n.a	n.a
Beryllium carbide	Be$_2$C, 30.035	1900	—	2100	21.0	1397	10.5×10^{-6}
Boron carbide	B$_4$C, 55.255	2512	4500	2350~2427	27	1854	2.63×10^{-6} ~ 5.6×10^{-6}
Chromium carbide	Cr$_7$C$_3$, 400.005	6992	109.0	1665	—	—	11.7×10^{-6}
Diamond	C, 12.011	3515.24	>10^{16} (types I and II a) >10^3 (type II b)	3550	900 (type I) 2400 (type II a)	—	2.16×10^{-6}
Graphite	C, 12.011	2250	1385	3650	—	—	0.6×10^{-6} ~ 4.3×10^{-6}
Hafnium monocarbide	HfC, 190.501	12670	45.0	3890~3950	22.15	—	6.3×10^{-6}
Lanthanum dicarbide	LaC$_2$, 162.928	5290	68.0	2360~2438	—	—	12.1×10^{-6}
Molybdenum monocarbide	MoC, 107.951	9159	50.0	2577	—	—	5.76×10^{-6}
Niobium monocarbide	NbC, 104.917	7820	51.1~74.0	3760	14.2	—	6.84×10^{-6}
Silicon monocarbide (moissanite)	α-SiC, 40.097	3160	4.1×10^5	2093 Transformation T	42.5	690	4.3×10^{-6} ~ 4.6×10^{-6}
Tantalum monocarbide	TaC, 194.955	14800	30~42.1	3880	22.2	190	6.64×10^{-6} ~ 8.4×10^{-6}
Thorium monocarbide	ThC, 244.089	10670	25.0	2621	28.9	—	6.48×10^{-6}
Titanium monocarbide	TiC, 59.878	4938	52.5	3140±90	17~21	—	7.5×10^{-6} ~ 7.7×10^{-6}

Continues Table 1.4

IUPAC name (synonyms name)	Theoretical chemical formula, relative molecular mass	Density ρ/kg·m^{-3}	Electrical resistivity $\rho/\mu\Omega\cdot cm$	Melting point /°C	Thermal conductivity κ/W·(m·K)$^{-1}$	Specific heat capacity c_p /J·(kg·K)$^{-1}$	Coefficient of linear thermal expansion α/K^{-1}
Tungsten monocarbide	WC, 195.851	15630	19.2	2870	121	—	6.9×10^{-6}
Uranium monocarbide	UC, 250.040	13630	50.0	2370~2790	23.0	—	11.4×10^{-6}
Vanadium monocarbide	VC, 62.953	5770	65.0~98.0	2810	24.8	—	4.9×10^{-6}
Zirconium monocarbide	ZrC, 103.235	6730	68.0	3540~3560	20.61	205	6.82×10^{-6}

Young's modulus E/GPa	Flexural strength τ/MPa	Compressive strength α/MPa	Vickers hardness H_V (Mohs H_M)	Other physicochemical properties, corrosion resistance, and uses
—	—	—	—	Decomposed in water with evolution of CH$_4$
314.4	—	723	2410H$_K$	Brick-red or yellowish-red octahedra. Used in nuclear-reactor cores
440~470	—	2900	3200~3500 H$_K$ (H$_M$9)	Hard, black, shiny crystals, the fourth hardest material known after diamond, cubic boron nitride, and boron oxide. It does not burn in an O$_2$ flame if the temperature is maintained below 983°C. Maximum operating temperature 2000°C (inert or reducing environment) or 600°C (oxidizing environment). It is not attacked by hot HF or chromic acid. Used as abrasive, and in crucibles for molten salts, except molten alkali-metal hydroxides. In the form of molded shapes, it is used for pressure blast nozzles, wire-drawing dies, and bearing surfaces for gauges. For grinding and lapping applications, the available mesh sizes cover the range 240 to 800
—	—	—	1336	Resists oxidation in the range 800~1000°C. Corroded by the following molten metals: Ni and Zn
930	—	7000	8000H$_K$ (H$_M$10)	Type I contains 0.1%~0.2% N, type IIa is N-free, and type IIb is very pure, generally blue in color. E-lectrical insulator ($E_g = 7eV$). Burns in oxygen
6.9	—	—	(H$_M$2)	High-temperature lubricant, used in crucibles for handling molten metals such as Mg, Al, Zn, Ga, Sb, and Bi
424	—	—	1870~2900	Dark, gray, brittle solid, the most refractory binary material known. Used in control rods in nuclear reactors and in crucibles for melting HfO$_2$ and other oxides. Corrosion-resistant to liquid metals such as Nb, Ta, Mo, and W. Severe oxidation in air above 1100~1400°C, but stable up to 2000°C in helium
—	—	—	—	Decomposed by H$_2$O

Continues Table 1.4

Young's modulus E/GPa	Flexural strength τ/MPa	Compressive strength α/MPa	Vickers hardness H_V (Mohs H_M)	Other physicochemical properties, corrosion resistance, and uses
197	—	—	1800 ($H_M > 9$)	Oxidized in air at 700~800℃
340	—	—	2470 ($H_M > 9$)	Lavender-gray powder, soluble in HF-HNO$_3$ mixture. Wear-resistant film, used for coating graphite in nuclear reactors. Oxidation in air becomes severe only above 1000℃
386~414	—	500	2400~2500 (H_M 9.2)	Semiconductor ($E_g = 3.03$ eV). Soluble in fused alkalimetal hydroxides
364	—	—	1599~1800 (H_M 9~10)	Golden-brown crystals, soluble in HF-HNO$_3$ mixture. Used in crucibles for melting ZrO$_2$ and similar oxides with high melting points. Corrosion-resistant to molten metals such as Ta, and Re. Readily corroded by liquid metals such as Nb, Mo, and Sn. Burning occurs in pure oxygen above 800℃. Severe oxidation in air above 1100~1400℃. Maximum operating temperature of 3760℃ under helium
—	—	—	1000	Readily hydrolyzed in water, evolving C$_2$H$_6$
310~462	—	1310	2620~3200 (H_M 9~10)	Gray crystals. Superconducting at 1.1K. Soluble in HNO$_3$ and aqua regia. Resistant to oxidation in air up to 450℃. Maximum operating temperature 3000℃ under helium. Used in crucibles for handling molten metals such as Na, Bi, Zn, Pb, Sn, Rb, and Cd. Corroded by the following liquid metals: Mg, Al, Si, Ti, Zr, V, Nb, Ta, Cr, Mo, Mn, Fe, Co, and Ni. Attacked by molten NaOH
421	—	—	3000	Black. Resistant to oxidation in air up to 700℃. Corrosion-resistant to Mo
—	—	—	3000	Corroded by molten Nb, Mo, and Ta
614	—	613	2090	Black crystals, soluble in HNO$_3$ with decomposition. Used in wear-resistant films and cutting tools. Resistant to oxidation in airup to 300℃
345	—	1641	1830~2930 ($H_M > 8$)	Dark gray, brittle solid, soluble in HF solutions containing nitrates or peroxide ions. Used in nuclear power reactors and in crucibles for handling molten metals such as Bi, Cd, Pb, Sn, and Rb, and molten zirconia (ZrO$_2$). Corroded by the following liquid metals: Mg, Al, Si, V, Nb, Ta, Cr, Mo, Mn, Fe, Co, Ni, and Zn. In air, oxidized rapidly above 500℃. Maximum operating temperature of 2350℃ under helium

1.4 Basic Properties of Advanced Ceramics

Table 1.5 Properties of carbides according to DIN EN 60672

Properties	Symbol	Units	SiC, sintered	SiC, silicon-infiltrated	SiC, recrystallized	SiC, nitride-bonded	Boron carbide
Open porosity	e	vol.%	0	0	0~15	—	0
Density, minimum		Mg/m^3	3.08~3.15	3.08~3.12	2.6~2.8	2.82	2.50
Bending strength	σ_B	MPa	300~600	180~450	80~120	200	400
Young's modulus	E	GPa	370~450	270~350	230~280	150~240	390~440
Hardness	H_V	100	25~26	14~25	25	—	30~40
Fracture toughness	K_{IC}	$MPa \cdot m^{1/2}$	3.0~4.8	3.0~5.0	3.0	—	3.2~3.6
Resistivity at 20 ℃	ρ_{20}	$\Omega \cdot m$	$10^3 \sim 10^4$	$2 \times (10^1 \sim 10^3)$	—	—	—
Resistivity at 600 ℃	ρ_{600}	$\Omega \cdot m$	10	5	—	—	—
Average coefficient of thermal expansion at 30~600 ℃	$\alpha_{30\sim1000}$	K^{-1}	$4 \times 10^{-6} \sim 4.8 \times 10^{-6}$	$4.3 \times 10^{-6} \sim 4.8 \times 10^{-6}$	4.8×10^{-6}	4.5×10^{-6}	6×10^{-6}
Specific heat at 30~100 ℃	$c_{p,30\sim1000}$	$J/(kg \cdot K)$	600~1000	650~1000	600~900	800~900	—
Thermal conductivity	$\lambda_{30\sim100}$	$W/(m \cdot K)$	40~120	110~160	20	14~15	28
Thermal fatigue resistance		Rated	Very good	Very good	Very good	Very good	—
Typical maximum application temperature	T	℃	1400~1750	1380	1600	1450	700~1000

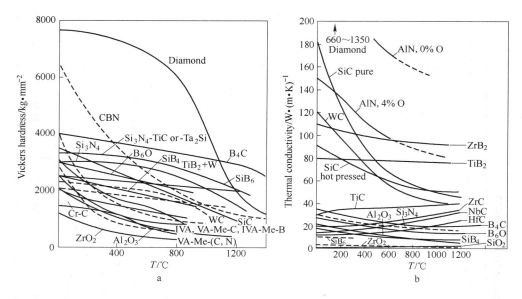

Figure 1.1 Temperature dependence of (a) the hardness and (b) the thermal conductivity of ceramic materials in comparison with diamond and cubic boron nitride

hardening constituents in all carbon-containing steels.

1.4.2.3 Nitrides

Some properties of nitrides are listed in Tables 1.6 and Table 1.7. Silicon nitride is the dominant nitride ceramic material because of its favorable combination of properties. The preparation of powders for the formation of dense silicon nitride materials requires the use of precursors. Four routes for the production of Si_3N_4 powders are used in practice: nitridation of silicon, chemical vapor deposition from $SiCl_4 + NH_3$, carbothermal reaction of SiO_2, and precipitation of silicon diimide $Si(NH)_2$ followed by decomposition. Table 1.8 gives examples of the properties of the resulting powders.

1.4.2.4 Silicides

Silicides, being compounds of silicon with metals, mostly show a metallic luster. Like intermetallic phases the silicides of a metal may occur in different stoichiometric variants, e.g. Ca_2Si, Ca_5Si_3, and $CaSi$. Silicides of the non-noble metals are unstable in contact with water and oxidizing media. Silicides of transition metals are highly oxidation-resistant. Silicides are used as ceramic materials mainly in high-temperature applications. $MoSi_2$ is used in resistive heating elements. Some properties are listed in Table 1.9.

Table 1.6 Properties of nitrides according to DIN EN 60672

Properties			Si_3N_4, sintered	Si_3N_4, reaction-bonded	Si_3N_4, hot-pressed	Aluminium nitride
Open porosity	e	vol. %	—	—	0	0
Density, minimum		g/cm^3	$3 \sim 3.3$	$1.9 \sim 2.5$	$3.2 \sim 3.4$	3.0
Bending strength	σ_B	MPa	$700 \sim 1000$	$200 \sim 330$	$600 \sim 800$	200
Young's modulus	E	GPa	$250 \sim 330$	$80 \sim 180$	$600 \sim 800$	320
Hardness	HV	100	$4 \sim 18$	$8 \sim 10$	$15 \sim 16$	11
Fracture toughness	K_{IC}	$MPa \cdot m^{1/2}$	$5 \sim 8.5$	$1.8 \sim 4.0$	$6.0 \sim 8.5$	3.0
Resistivity at 20 °C	ρ_{20}	$\Omega \cdot m$	10^{11}	10^{13}	10^{13}	10^{13}
Resistivity at 600 °C	ρ_{600}	$\Omega \cdot m$	10^2	10^{10}	10^9	10^{12}
Average coefficient of thermal expansion at 30 $\sim$ 600 °C	$\alpha_{30 \sim 1000}$	K^{-1}	$2.5 \times 10^{-6} \sim 3.5 \times 10^{-6}$	$2.1 \times 10^{-6} \sim 3 \times 10^{-6}$	$3.1 \times 10^{-6} \sim 3.3 \times 10^{-6}$	$4.5 \times 10^{-6} \sim 5 \times 10^{-6}$
Specific heat capacity at 30 $\sim$ 100°C	$c_{p, 30 \sim 1000}$	$J \cdot (kg \cdot K)^{-1}$	$700 \sim 850$	$700 \sim 850$	$700 \sim 850$	—
Thermal conductivity	$\lambda_{30 \sim 100}$	$W \cdot (m \cdot K)^{-1}$	$15 \sim 45$	$4 \sim 15$	$15 \sim 40$	> 100
Thermal fatigue resistance		Rated	Very good	Very good	Very good	Very good
Typical maximum application temperature	T	°C	1250	1100	1400	—

Table 1.7 Physical properties of nitrides and nitride-based high-temperature ceramics

IUPAC name (synonyms name)	Theoretical chemical formula, relative molecular mass	Density $\rho/\text{kg} \cdot \text{m}^{-3}$	Electrical resistivity $\rho/\mu\Omega \cdot \text{cm}$	Melting point /℃	Thermal conductivity $\kappa/\text{W} \cdot (\text{m} \cdot \text{K})^{-1}$	Specific heat capacity c_p /J·(kg·K)$^{-1}$	Coefficient of linear thermal expansion α/K^{-1}
Aluminium mononitride	AlN, 40.989	3050	10^{17}	2230	29.96	820	5.3×10^{-6}
Boron mononitride	BN, 24.818	2250	10^{19}	2730(dec.)	15.41	711	7.54×10^{-6}
Chromium mononitride	CrN, 66.003	6140	640	1499(dec.)	12.1	795	2.34×10^{-6}
Hafnium mononitride	HfN, 192.497	13840	33	3310	21.6	210	6.5×10^{-6}
Molybdenum mononitride	MoN, 109.947	9180	—	1749	—	—	—
Niobium mononitride	NbN, 106.913	8470	78	2575	3.63	—	10.1×10^{-6}
Silicon nitride	β-Si$_3$N$_4$, 140.284	3170	10^6	1850	28	713	2.25×10^{-6}
Silicon nitride	α-Si$_3$N$_4$, 140.284	3184	10^{19}	1900(sub.)	17	700	$2.5 \times 10^{-6} \sim 3.3 \times 10^{-6}$
Tantalum mononitride	TaN, 194.955	13800	128~135	3093	8.31	210	3.2×10^{-6}
Thorium mononitride	ThN, 246.045	11560	20	2820	—	—	7.38×10^{-6}
Titanium mononitride	TiN, 61.874	5430	21.7	2930(dec.)	29.1	586	9.35×10^{-6}
Tungsten mononitride	WN	15940	—	593	—	—	—
Uranium mononitride	UN, 252.096	14320	208	2900	12.5	188	9.72×10^{-6}
Vanadium mononitride	VN, 64.949	6102	86	2360	11.25	586	8.1×10^{-6}
Zirconium mononitride	ZrN, 105.231	7349	13.6	2980	20.90	377	7.24×10^{-6}

Young's modulus E/GPa	Flexural strength τ/MPa	Compressive strength σ/MPa	Vickers hardness H_V (Mohs H_M)	Other physicochemical properties, corrosion resistance, and uses
346	—	2068	1200 (H_M 9~10)	Insulator (E_g = 4.26eV). Decomposed by water, acids, and alkalis to Al(OH)$_3$ and NH$_3$. Used in crucibles for GaAs crystal growth

Continues Table 1.7

Young's modulus E/GPa	Flexural strength τ/MPa	Compressive strength σ/MPa	Vickers hardness H_V ($Mohs\ H_M$)	Other physicochemical properties, corrosion resistance, and uses
85.5	—	310	230 ($H_M 2.0$)	Insulator ($E_g = 7.5\text{eV}$). Used in crucibles for molten metals such as Na, B, Fe, Ni, Al, Si, Cu, Mg, Zn, In, Bi, Rb, Cd, Ge, and Sn. Corroded by these molten metals: U, Pt, V, Ce, Be, Mo, Mn, Cr, V, and Al. Attacked by the following molten salts: PbO_2, Sb_2O_3, Bi_2O_3, KOH, and K_2CO_3. Used in furnace insulation, diffusion masks, and passivation layers
—	—	—	1090	
—	—	—	1640 ($H_M 8\sim 9$)	Most refractory of all nitrides
—	—	—	650	
55	—	—	1400 ($H_M > 8$)	Dark gray crystals. Transition temperature 15.2K. Insoluble in HCl, HNO_3, and H_2SO_4, but attacked by hot caustic solutions, lime, or strong alkalis, evolving NH_3
—	—	—	($H_M > 9$)	
304	—	—	($H_M > 9$)	Gray amorphous powder or crystals. Corrosion-resistant to molten metals such as Al, Pb, Zn, Cd, Bi, Rb, and Sn, and molten salts NaCl-KCl, NaF, and silicate glasses. Corroded by molten Mg, Ti, V, Cr, Fe, Co, cryolite, KOH, and Na_2O
—	—	—	1110 ($H_M > 8$)	Bronze-colored or black crystals. Transition temperature 1.8K. Insoluble in water, slowly attacked by aqua regia, HF, and HNO_3
—	—	—	600	Gray solid. Slowly hydrolyzed by water
248	—	972	1900 ($H_M 8\sim 9$)	Bronze-colored powder. Transition temperature 4.2K. Corrosion-resistant to molten metals such as Al, Pb, Mg, Zn, Cd, and Bi. Corroded by molten Na, Rb, Ti, V, Cr, Mn, Sn, Ni, Cu, Fe, and Co. Dissolved by boiling aqua regia; decomposed by boiling alkalis, evolving NH_3
—	—	—	—	
149	—	—	455	Gray solid. Slowly hydrolyzed by water

Continues Table 1.7

Young's modulus E/GPa	Flexural strength τ/MPa	Compressive strength σ/MPa	Vickers hardness H_V (Mohs H_M)	Other physicochemical properties, corrosion resistance, and uses
—	—	—	1520 (HM 9~10)	Black powder. Transition temperature 7.5K. Soluble in aqua regia
—	—	979	1480 (H_M >8)	Yellow solid. Transition temperature 9K. Corrosion-resistant to steel, basic slag, and cryolite, and molten metals such as Al, Pb, Mg, Zn, Cd, and Bi. Corroded by molten Be, Na, Rb, Ti, V, Cr, Mn, Sn, Ni, Cu, Fe, and Co. Soluble in concentrated HF, slowly soluble in hot H_2SO_4

Table 1.8 Characteristics of Si_3N_4 powders processed by different preparation techniques (E, equiaxed; R, rod-like)

Properties, compositions and crystallinity	Technique							
	Nitridation of Si		Chemical vapor deposition		Carbothermal reduction		Diimide precipitation	
Sample no.	1	2	1	2	1	2	1	2
Specific surface area/$m^2 \cdot g^{-1}$	23	10	11	4	10		11	13
O/wt%	1.4	1.0	1.0	3.0	2.0		1.4	1.5
C/wt%	0.2	0.25	—	—	0.9		0.1	0.1
Fe, Al, Ca/wt%	0.07	0.4	0.005	0.005	0.22		0.01	0.015
Other impurities/wt%			Cl 0.04,	Mo + Ti 0.02			Cl 0.1	Cl 0.005
Crystallinity/%	100	100	60	0	100		98	—
$\alpha/(\alpha+\beta)$/%	95	92	95	—	98		86	95
Morphology[①]	E	E	E+R	E+R	E+R		E	E

① E—equiaxed; R—rod-like.

Table 1.9 Physical properties of silicides and silicide-based high-temperature ceramics

IUPAC name (synonyms names)	Theoretical chemical formula, relative molecular mass	Density $\rho/\mathrm{kg}\cdot\mathrm{m}^{-3}$	Electrical resistivity $\rho/\mu\Omega\cdot\mathrm{cm}$	Melting point /°C	Thermal conductivity $\kappa/\mathrm{W}\cdot(\mathrm{m}\cdot\mathrm{K})^{-1}$	Specific heat capacity c_p /$\mathrm{J}\cdot(\mathrm{kg}\cdot\mathrm{K})^{-1}$	Coefficient of linear thermal expansion α/K^{-1}
Chromium disilicide	CrSi$_2$, 108.167	4910	1400	1490	106	—	13.0×10^{-6}
Hafnium disilicide	HfSi$_2$, 234.66	8030	—	1699	—	—	—
Molybdenum disilicide	MoSi$_2$, 152.11	6260	21.5	1870	58.9	—	8.12×10^{-6}
Niobium disilicide	NbSi$_2$, 149.77	5290	50.4	2160	—	—	—
Tantalum disilicide	TaSi$_2$, 237.119	9140	8.5	2299	—	—	$8.8\times10^{-6}\sim 9.54\times10^{-6}$
Thorium disilicide	ThSi$_2$, 288.209	7790	—	1850	—	—	—
Titanium disilicide	TiSi$_2$, 104.051	4150	123	1499	—	—	10.4×10^{-6}
Tungsten disilicide	WSi$_2$, 240.01	9870	33.4	2165	—	—	8.12×10^{-6}
Uranium disilicide	USi$_2$, 294.200	9250	—	1700	—	—	—
Vanadium disilicide	VSi$_2$, 107.112	5100	9.5	1699	—	—	11.2×10^{-6}
Zirconium disilicide	ZrSi$_2$, 147.395	4880	161	1604	—	—	8.6×10^{-6}

IUPAC name	Vickers hardness H_V (Mohs H_M)	Compressive strength σ/MPa	Flexural strength τ/MPa	Young's modulus E/GPa	Other physicochemical properties, corrosion resistance, and uses
Chromium disilicide	1000~1130	—	—	—	The compound is thermally stable in air up to 1000 °C. Corrosion-resistant to molten metals such as Zn, Pd, Ag, Bi, and Rb. It is corroded by the following liquid metals: Mg, Al, Si, V, Cr, Mn, Fe, Ni, Cu, Mo, and Ce
Hafnium disilicide	865~930	—	—	—	
Molybdenum disilicide	1260	2068~2415	—	407	
Niobium disilicide	1050	—	—	—	Corroded by molten Ni
Tantalum disilicide	1200~1600	—	—	—	Corrosion-resistant to molten Cu, while corroded by molten Ni
Thorium disilicide	1120	—	—	—	
Titanium disilicide	890~1039	—	—	—	
Tungsten disilicide	1090	—	—	—	
Uranium disilicide	700	—	—	—	
Vanadium disilicide	1400	—	—	—	Corrosion-resistant to molten Cu, while corroded by molten Ni
Zirconium disilicide	1030~1060	—	—	—	

1.5 Applications of Advanced Ceramics

The traditional ceramic industries continue to be necessary to our modern society, but whole new categories of specialty applications have arisen. For many of these applications, the ceramics have been carefully engineered to meet demanding and often narrow specification of properties and configurations. Excellent properties, advanced ceramics may be found in aircraft engines[6], automotive engines[7], cutting tools used for making metal products[8,9], the skin of space shuttles[10], knives, bullet proof armor[11], artificial hip-joints, computers[12], microelectronics[13] and so on[14]. R. Waser[15] drawn a chart of application field of advanced ceramics as shown in Figure 1.2.

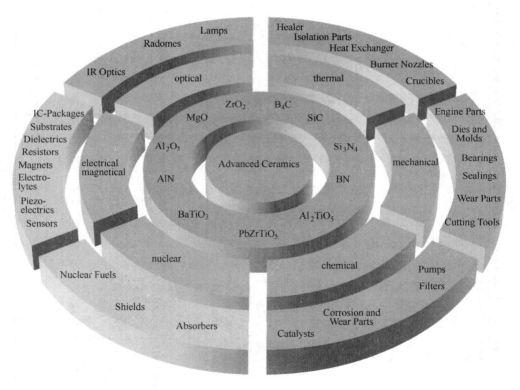

Figure 1.2 Functional and structural applications of advanced ceramics[15]

The ceramics industry is a very large international industry, the technology of ceramics is a rapidly developing applied science in today's world. Technological advances result from unexpected material discoveries. On the other hand, the new technology can drive the development of new ceramics. Currently many new classes of materials have been devised to satisfy various new applications. In most of these applications, improved materials based on ceramics were purposefully sought after. These applications dramatically change or affect the environment in which we live. Not only do we have economic and material issues to deal with, but also, unforeseeable changes in economic factors and the political environment will play significant roles in the needs for improved components and devices, as well as affect our ability to apply resources towards research and development needed to bring new materials to the marketplace. Ceramics application could be categorized as structural ceramics, electrical ceramics, ceramic composites, and ceramic coatings. These materials are emerging as the leading class of materials needed to be improved to explore further potential applications. Current and future advanced ceramic products are indicated in Table 1.10.

Table 1.10 Current and future products for advanced ceramics[16~18]

Mechanical engineering	Aerospace	Automotive	Defense industry
Cutting tools and dies	Fuel systems and valves	Heat engines	Tank power train
Abrasives	Power units	Calalytic converters	Submarine shaft seals
Precise instrument parts	Low weight components	Drivetrain components	Improved armors
Molten metal filter	Fuel cells	Turbines	Propulsion systems
Turbine engine components	Thermal protection systems	Fixed boundary recuperators	Ground support vehicles
Low weight components for rotary equipment	Turbine engine components	Fuel injection components	Military weapon systems
Wearing parts	Combustors	Turbocharger rotors	Millitary aircraft
Bearings	Bearings	Low heat rejection diesels	Wear-resistant precision bearings
Seals	Seals	Waterpump seals	
Solid lubricants	Structures		
Biological, chemical processing engineering	Electrical, magnetic engineering	Nuclear industry	
Artificial teeth, bones and joins	Memory elements	Nuclear fuel	
Catalysts and igniters	Resistance heating element	Nuclear fuel cladding	

Continues Table 1.10

Biological, chemical processing engineering	Electrical, magnetic engineering	Nuclear industry	
Heart valves	Varistor sensor	Control materials	
Heat exchanger	Intergrated circuit substrate	Moderating materials	
Reformers	Multilayer capacitors	Reactor mining	
Recuperators	Advanced multilayer integrated packages		
Refractories			
Nozzles			
Oil industry	Electric power generation	Optical engineering	Thermal engineering
Bearings	Bearings	Laser diode	Electrode materials
Flow control valves	Ceramic gas turbines	Optical communication cable	Heat sink for electronic parts
Pumps	High temperature components	Heat resistant translucent porcelain	High-temperature industrial furnace lining
Refinery heater	Fuel cells	Light emitting diode	
Blast sleeves	Filters		

Today, advanced ceramics have been widely used in wearing parts, seals, low weight components and fuel cells in transportation sectors, to reduce the weight of product, increase performance especially at high temperatures, prolong the life cycle of a product and improve the efficiency of combustion. As advances in ceramic technology offer potential and immediate opportunities, these materials will translate into greater market shares in transportation sectors.

In 2009, a demand for advanced ceramics to advance 3.8% per year to $12.6 billion in 2012 is forecast, as shown in Table 1.11.

Table 1.11 Advanced ceramics demand by market[19]   (million dollars)

Year	1997	2002	2007	2012	2017
Electronic components	2763	2394	2620	3020	3385

Continues Table 1.11

Year	1997	2002	2007	2012	2017
Electrical equipment	1217	1349	1880	2395	2940
Industrial machinery	914	965	1410	1780	2280
Transportation equipment	840	992	1325	1735	2255
Chemical and plastic	606	713	1130	1380	1665
Body armor	3	30	650	350	350
Environmental	319	420	575	755	980
Medical products	60	96	155	240	360
Other markets	378	506	715	975	1285
Total market	**7100**	**7465**	**10460**	**12630**	**15500**

Jürgen Rödel, et al[20] have recently developed a roadmap for advanced ceramics for the period from 2010 to 2025 to provide guidelines for future investments for policy makers, scientists and industry alike. Based on questionnaires, interviews and a final workshop with well-balanced participation of members from industry and academia three roadmaps on application fields and two roadmaps on scientific areas have been developed and contrasted. The three application fields selected are: (1) electronics, information and communication; (2) energy and environment; (3) mechanical engineering and the two scientific fields are: 1) structural and functional properties; 2) process technology. Within these fields the tremendous growth opportunities for ceramics as an enabling technology are highlighted and manifold suggestions for future development are provided.

Study Guide

1-1　What is a ceramic?

1-2　How to classify the advanced ceramics by their main function?

1-3　List some important applications of ceramics in an automobile.

References

[1] Carter C B, Norton M G. Ceramic Materials: Science and Engineering [M]. New York: Springer-Verlag New York Inc, 2007 (4): 740.

[2] Padgett G C. Classification and Standardization of Advanced Technical Ceramics [J]. Key Engineering Materials, 1991, 56: 411~418.

[3] Fritz Aldinger, Nils Claussen, Richard M. Spriggs, et al. Handbook of Advanced Ceramics [M]. Elsevier Inc., 2003.

[4] Thümmler F. Engineering Ceramics [J]. Journal of the European Ceramic Society, 1990, 6 (3): 139~151.

[5] Greg Perry. Advanced Technical Ceramics-A Comprehensive History and Guide [J]. Reference and Education: Science, 2010: 6.

[6] Backman D G, Williams J C. Advanced Materials for Aircraft Engine Applications [J]. Science, 1992, 255 (5048): 1082~1087.

[7] Matsui M. Advanced Technical Ceramics: Their Potential [J]. Ceramics International, 1993, 19 (1): 9~16.

[8] Li X S, Low I. Ceramic Cutting Tools-An Introduction [J]. Key Engineering Materials, 1994, 96: 1~18.

[9] Xikun L, Jing L, Like Q, et al. Composition, Characteristics and Development of Advanced Ceramic Cutting Tools [J]. Journal of Rare Earths, 2007, 25: 287~294.

[10] Schmidt S, Beyer S, Knabe H, et al. Advanced Ceramic Matrix Composite Materials for Current and Future Propulsion Technology Applications. Acta Astronautica, 2004 (55): 409~420.

[11] Matchen B. Applications of Ceramics in Armor Products [J]. Key Engineering Materials, 1996, 122: 333~344.

[12] Terrell A. Vanderah. Talking ceramics [J]. Science, New Series, 2002, 298 (5596): 1182~1184.

[13] Roy R. Ceramic Packaging for Electronics [J]. Key Engineering Materials, 1996, 122: 17~34.

[14] Cheetham A K. Advanced Inorganic Materials: an Open Horizon [J]. Science (New York, NY), 1994, 264 (5160): 794~795.

[15] Waser R. Advanced Ceramic Materials: Basic Research Viewpoint, 27. http: //202.193. 70. 166/mirror/MATEMIRROR/materialmirror 13/MATEA04806 – 759547. pdf.

[16] Evans A G. Ceramics and Ceramic Composites as High-Temperature Structural Materials: Challenges and Opportunities [and Discussion] [J]. Philosophical Transactions of the Royal Society of London. Series A: Physical and Engineering Sciences, 1995, 351 (1697): 511~527.

[17] Bardhan P. Ceramic Honeycomb Filters and Catalysts [J]. Current Opinion in Solid State and Materials Science, 1997, 2 (5): 577~583.

[18] Sommers A, Wang Q, Han X, et al. Ceramics and Ceramic Matrix Composites for Heat Exchangers in Advanced Thermal Systems-A Review [J]. Applied Thermal Engineering, 2010, 30 (11): 1277~1291.

[19] Marvin D. Sustainability Defined [J]. Ceramic Industry, 2009, 159 (7): 13.

[20] Rödel J, Kounga A B N, Weissenberger-Eibl M, et al. Development of a Roadmap for Advanced Ceramics: 2010~2025 [J]. Journal of the European Ceramic Society, 2009, 29 (9): 1549~1560.

2 Powder Processing

The nature of the raw material has a decisive effect on the properties of final product. Purity, particle size distribution, reactivity, polymorphic form, availability, and cost must all be considered and carefully controlled. In this chapter we discuss the types and sources of raw materials and the preparation processing of advanced ceramic powders.

2.1 Raw Materials

Raw material is very important in ceramics in various aspects. It is one of the most decisive factors for the quality of a product. Use of proper raw material is critical for producing ceramics of high quality at a minimal cost. Technologically, the raw powder sets a fair starting point for industries[1].

During the past 100 years scientists and engineers have acquired a much better understanding of ceramic materials and their processing and have found that naturally occurring minerals could be refined or new compositions synthesized to achieve unique properties. These refined or new synthesized ceramic materials typically are of highly controlled composition and structure and have been engineered to fill the needs of applications too demanding for traditional ceramics.

2.1.1 Oxides

The raw materials used for oxide ceramics are almost entirely produced by chemical processes to achieve a high chemical purity and to obtain the most suitable powders for product fabrication. Here we discuss individually some important oxides.

2.1.1.1 Alumina

Alumina (Al_2O_3) has excellent physical and chemical properties, and is used in a variety of industrial fields. Various types of high-purity grades of powders such as hydrated, activated, calcined, low-soda, reactive, tabular and fused aluminas are available in the market.

Aluminum oxide occurs naturally as the mineral corundum, which is better known to most of us when it is in gem-quality crystals called ruby and sapphire. Ruby and sapphire are precious gems, because of their chemical inertness and hardness. Al_2O_3 powder is produced in large quantity from the mineral bauxite by the Bayer process. Bauxite is primarily colloidal aluminum hydroxide intimately mixed with iron hydroxide and other impurities. In the Bayer process, the Al_2O_3 component in bauxite dissolves in NaOH solution to form $NaAlO_2$, subsequently followed by the precipitation of $Al(OH)_3$ by the hydrolysis of $NaAlO_2 \cdot Al_2O_3$ can be produced by calcining $Al(OH)_3$ at above 1000℃. Through the process, impurities (e.g. SiO_2, Fe_2O_3 and TiO_2) are removed and a nominal 99.5% Al_2O_3 can be produced with Na_2O (0.5% ~ 0.05%) as the dominant impurity. Low-soda calcined alumina contains Na_2O of below 0.1% and is applied to spark plugs, substrates, electronic components, and various engineering ceramics parts. Submicron-sized alumina powders with purity 99.8% ~ 99.9% are available. High purity (99.99% or more) and ultra-fine powders are also synthesized through advanced processes: hydrolysis and heat treatment of aluminum alkoxide, and pyrolysis of ammonium alum, $(NH_4)_2SO_4 \cdot Al_2(SO_4)_3 \cdot 12H_2O$, or ammonium dawsonite, $NH_4AlCO_3(OH)_2$, through the following reactions:

(1) Hydrolysis and heat treatment of aluminum alkoxide:

$$2Al + 6ROH \longrightarrow 2Al(OR)_3 + 3H_2$$
$$2Al(OR)_3 + 4H_2O \longrightarrow Al_2O_3 \cdot H_2O + 6ROH$$
$$Al_2O_3\text{-}H_2O \longrightarrow \text{Calcination} \longrightarrow Al_2O_3 + H_2O$$

(2) Pyrolysis of ammonium alum:

$$Al_2(SO_4)_3 \cdot 18H_2O + (NH_4)_2SO_4 \longrightarrow (NH_4)_2SO_4 \cdot Al_2(SO_4)_3 \cdot 12H_2O + 6H_2O$$
$$(NH_4)_2SO_4 \cdot Al_2(SO_4)_3 \cdot 12H_2O \longrightarrow \text{Pyrolysis} \longrightarrow Al_2O_3 + 2NH_3 + 4SO_3 + 13H_2O$$

(3) Pyrolysis of ammonium dawsonite:

$$(\text{Aluminum salt}) + NH_4HCO_3 \longrightarrow NH_4AlCO_3(OH)_2$$
$$2NH_4AlCO_3(OH)_2 \longrightarrow Al_2O_3 + 2NH_3 + 2CO_2 + 3H_2O$$

In these production methods, aluminum intermediate resources made from the Bayer process are used as a starting material.

Low-soda aluminas have soda levels of less than 0.1%. These aluminas are calcined alumina, from which the residual soda had been removed to make it better suited to ceramic processing. Reduction of the soda content is achieved by adding boric acid or silica to aluminum trihydroxide during the calcination process. Low-soda alumina is widely used to help in sintering of electronic components and various structural ceramics. Low-soda alumina is also used for abrasives and polishing compounds. Figure 2.1 shows the SEM images of low-soda aluminas produced by Jingang New Materails Co. (China) with a 0.2% addition of boron acid and 0.1% addition of MgO respectively calcining at 1350℃ for 2 h. Both alumina powders have a 0.02% content of Na_2O.

Al_2O_3 powder is used in the manufacture of porcelain, alumina laboratory ware, cru-

Figure 2.1 SEM images of Al_2O_3 powders
a—0.2% addition of boron acid; b—0.1% addition of MgO

cibles and metal casting molds, high-temperature cements, wear-resistant parts (sleeves, tiles, seals, etc), sandblast nozzles, armor, medical components, abrasives, refractories, and a variety of other components.

2.1.1.2 Magnesia

Magnesia (MgO) has a high melting point, 2800℃, and is an abundant natural resource. It reacts, however, with water, carbon dioxide, and acids, turning into magnesium hydroxide, magnesium carbonate, and other magnesium salts. Magnesia, unlike alumina or silica, does not have very stable chemical properties. If magnesium hydroxide, which is the raw ingredient for MgO, is heated gradually, fine crystals grow within the magnesium hydroxide. The specific surface of magnesium oxide crystals is large, several hundred square millimeters per gram, and it is strongly reactive; MgO is obtained by its decomposition. If the MgO is heated further, its activity is reduced rapidly but even at 1200℃ considerable activity remains. This highly active MgO is called active magnesia or light magnesia. For a better sintered body, pressure forming in this state combined with high-temperature sintering is the methods chosen, as in sintering other materials.

Magnesia clinker, used as a refractory material, is sold commercially with a density at 96% of theoretical. After products molded of magnesia clinker or light magnesia are melted in an electric arc furnace, recrystallized, electromelted magnesia is formed. The ceramic characteristics of MgO vary considerably, depending upon the type of mother salt and the calcination temperature. Therefore, these factors make selection of the raw material a critical point.

2.1.1.3 Zirconia

Pure zirconia (ZrO_2) has a high melting temperature (2700℃) and a low thermal conductivity. Its application includes raw materials for leadzirconia-titanate electronic ceramics, solid electrolyte, structural components and refractories. The crystal structure of pure zirconia is monoclinic at room temperature, transforming to tetragonal at 1170℃ and cubic at about 2370℃. The transformation of the monoclinic structure is accompanied by a large volume change and causes many cracks within the sintered structures.

Zirconia can be stabilized by the addition of some oxides called stabilizers, such as CaO, MgO and Y_2O_3. They form a solid solution with zirconia and make the cubic and/or tetragonal structure stable for all temperatures, eliminating the detrimental phase transformation during heating and cooling. Addition of more than 16% (mole fraction) of CaO, 16% (mole fraction) of MgO or 8% (mole fraction) of Y_2O_3 into the zirconia structure is needed to form a fully stabilized zirconia. This solid solution is termed as **stabilized zirconia**. Stabilized zirconia, which is a solid electrolyte, has high oxide ion conductivity and is used in applications such as electrodes, oxygen sensors and solid oxide fuel cells. **Partially stabilized zirconia** is obtained with a smaller addition of stabilizer. In this material, very fine precipitates of metastable tetragonal and/or monoclinic phases are dispersed in a cubic zirconia matrix. Typical amounts of additives to form partially stabilized zirconia are more than 8% (mole fraction) for MgO, 8% (mole fraction) for CaO or 3%~4% (mole fraction) for Y_2O_3. Partially stabilized zirconia has excellent fracture strength and high toughness at room temperature. It can be used for various structural applications such as grinding media, extrusion dies, ceramic liners, optical fiber connectors and medical implants. The toughening mechanisms in partially stabilized zirconia are microcracking and induced stress.

Zirconia powders are manufactured by electric fusion processes (90%~99% purity), chemical precipitation processes (99%~99.9% purity), such as co-precipitation, hydrolysis of zirconium salts or alkoxides, and hydrothermal processes. In fusion processes, zircon and carbon are heated in an electric arc furnace to above 2000℃. Silica (SiO_2) is reduced to SiO and evaporates leaving behind zirconia. The compounds containing stabilizer, such as $CaCO_3$, are added in the melt when stabilized zirconia are produced. Chemical precipitation processes are commonly used in the production of sinterable, fine and high purity zirconia powders. A typical process for the production of zirconia powder through the precipitation process is as follows:

$$\text{Zircon}(\text{ZrSiO}_4) + \text{NaOH} \xrightarrow{\text{Solution}} \text{Na}_2\text{ZrO}_3 \xrightarrow{\text{HCl}}$$
$$\text{ZrOCl}_2 \cdot 8\text{H}_2\text{O} \xrightarrow{\text{NH}_4\text{OH}} \text{Zr(OH)} \xrightarrow{\text{Calcination}} \text{ZrO}_2$$

Zircon is first converted to zirconium oxychloride ($\text{ZrOCl}_2 \cdot 8\text{H}_2\text{O}$) through the solution-precipitation process. Zirconium hydroxide, Zr(OH)_4, is precipitated by the addition of NH_4OH solution into zirconium oxychloride solution. After filteration and washing, the precipitated Zr(OH)_4 is calcined to form zirconia. Powder characteristics including grain size, particle shape, agglomerate size, and specific surface area can be adjusted within a certain degree by controlling the conditions of precipitation and calcination.

Stabilized or partially stabilized zirconia can be obtained by the co-precipitation process. During the chemical processing, a specific amount of the salt for stabilizers such as YCl_3 can be added into the zirconium oxychloride solution and mixed. Addition of NH_4OH solution into the mixed solution results in co-precipitation of Zr(OH)_4 and the hydroxides of stabilizer, such as Y(OH)_3, as mixtures. A cubic or tetragonal phase zirconia can be formed during calcination of the co-precipitated mixture. The powders obtained have chemically high uniformity and can be used in applications such as refractories, engineering ceramics and thermal barrier coatings.

The hydrolysis process is used to produce ZrO_2 or (ZrO_2 + oxides of stabilizer) powders. The precipitated sol is prepared through the hydrolysis of the solution of zirconium oxychloride and/or its mixture with the salt of stabilizers, such as YCl_3, with heating, and then calcined.

2.1.1.4 Zincite

Zinc oxide (ZnO) occurs naturally as the mineral zincite. Chemically pure ZnO is white. Zincite is red because it contains up to 10% Mn; traces of FeO are usually also present. Naturally occurring sources of zincite are not commercially important. There are two production methods for forming zinc oxide:

(1) Oxidation of vaporized zinc metal in air;
(2) Reduction of sphalerite (ZnS) with carbon and CO.

Sphalerite is a naturally occurring mineral and the most important ore of zinc. Large deposits are found in limestone of the Mississippi Valley, around Joplin, MO and Galena, IL. Significant deposits are also found in France, Mexico, Spain, Sweden, and the UK.

The largest consumers of ZnO are the rubber and adhesives industries. Zinc oxide is also found in some latex paints, tiles, glazes, and porcelain enamels, and is the most widely used material in the manufacture of vanstors.

2.1.1.5 Rutile and anatase

Rutile (TiO_2, titania) occurs as a constituent of igneous rocks such as granites and also in metamorphic derivatives such as gneiss. It also occurs as fine needles in slates, biotite mica, quartz, and feldspar. Economically the most important deposits are segregations in igneous rocks as found in Virginia, Canada, and Norway. Rutile is also an important constituent of beach sands resulting from denudation of rutile-bearing rocks, as in Australia, Florida, and India.

Titania is also produced by reacting ilmenite $FeTiO_3$ with sulfuric acid at 150 ~ 180℃ to form titanyl sulfate, $TiOSO_4$:

$$FeTiO_3(s) + 2H_2SO_4(aq) + 5H_2O(l) \longrightarrow FeSO_4 \cdot 7H_2O(s) + TiOSO_4(aq)$$

Titanyl sulfate is soluble in water and can be separated from undissolved impurities and the precipitated iron sulfate by filtration. Hydrolyzing at 90℃ causes the hydroxide $TiO(OH)_2$ to precipitate:

$$TiOSO_4(aq) + 2H_2O(l) \longrightarrow TiO(OH)_2(s) + H_2SO_4(aq)$$

The titanyl hydroxide is calcined at about 1000℃ to produce titania TiO_2.

2.1.1.6 Mullite

Mullite is a solid solution phase of alumina and silica commonly found in ceramics[2]. Only rarely does mullite occur as a natural mineral. According to introductory remarks made by Schneider and MacKenzie at the conference "Mullite 2000" (H. Schneider and K. MacKenzie, *J. Eur. Ceram. Soc.* **21**, Ⅲ (2001)), the geologists Anderson, Wilson, and Tait of the Scottish Branch of His Majesty's Geological Survey discovered the mineral mullite less than a century ago. The trio was collecting mineral specimens from ancient lava flows on the island of Mull off the west coast of Scotland when they chanced upon the first known natural deposit of this ceramic material. The specimens were initially identified as sillimanite, but later classified as mullite.

Being the only stable intermediate phase in the Al_2O_3-SiO_2 system at atmospheric pressure, mullite is one of the most important ceramic materials. The chemical formula for mullite is deceptively simple: $3Al_2O_3 \cdot 2SiO_2$.

In 1987, Klug et al[3] published their SiO_2-Al_2O_3 phase diagram as shown in Figure 2.2.

Mullite was first identified as the product of heating kaolinitic clays, resulting in a compound with an approximate alumina-to-silica molar ratio of 3 : 2. The order of reaction proceeds as follows:

$$2Al_2Si_2O_5(OH)_4 \xrightarrow{450℃} 2(Al_2O_3 \cdot 2SiO_2) + 4H_2O$$
Kaolinite → Metakaolin

$$2(Al_2O_3 \cdot 2SiO_2) \xrightarrow{925℃} 2Al_2O_3 \cdot 3SiO_2 + SiO_2$$
Metakaolin → Silicon spinel

$$2Al_2O_3 \cdot 3SiO_2 \xrightarrow{1100℃} 2(Al_2O_3 \cdot SiO_2) + SiO_2$$
Silicon spinel → Pseudomullite

$$3(Al_2O_3 \cdot SiO_2) \xrightarrow{1400℃} 3Al_2O_3 \cdot 2SiO_2 + SiO_2$$
Pseudomullite → Mullite + cristobalite

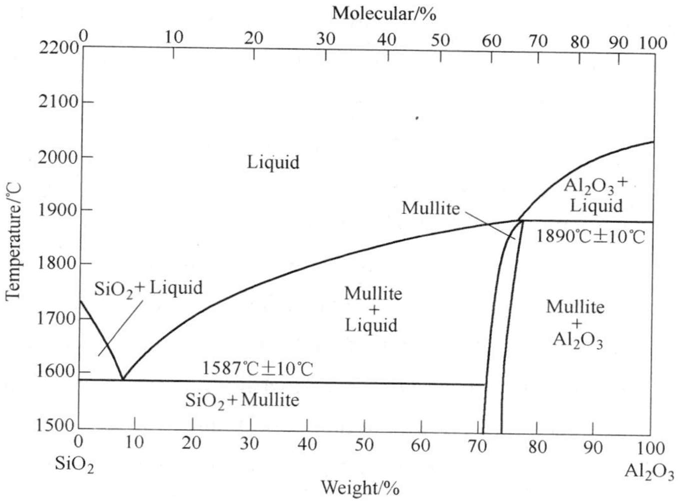

Figure 2.2 Phase diagram for the alumina-silica system[3]

Mullite has been fabricated into transparent, translucent, and opaque bulk forms. These materials may have optical and electronic device applications. Mullite's temperature stability and refractory nature are superior to corundum's in certain high-temperature structural applications. Another characteristic of this aluminosilicate is its temperature-stable defect structure, which may indicate a potential use in fuel cell electrolytes.

2.1.2 Nonoxides

Most of the important nonoxide raw materials do not occur naturally and therefore must be synthesized. The synthesis route is usually one of the following:

(1) Combine the metal directly with the nonmetal at high temperature.

(2) Reduce the oxide with carbon at high temperature (carbon thermal reduction) and subsequently react it with the nonmetal.

In this section, we look at several important nonoxide materials. To show the variety of nonoxide materials we have taken examples of carbides, nitrides, and brides.

2.1.2.1 Silicon carbide

Silicon carbide (SiC), an artificial material with a strong covalent bonding structure, was accidentally synthesized by E. G. Acheson in 1891. It is harder and more heat resistant than Al_2O_3, and it also demonstrates outstanding resistance to corrosion. Therefore, it is widely used as both an abrasive and a refractory material. It is also used as a metallurgical additive in iron-and steel making, and putting to use its electrical properties, in heating elements and in parts for electrical circuits (varistors and arrestors). The production method uses an electric furnace, basically following the process Acheson discovered. That is the only method of mass production.

Silicon carbide has two crystal forms α-SiC, which belongs to either a hexagonal or a rhombohedral phase, and β-SiC, which belongs to a cubic phase. Polytypes of silicon carbide are represented by the way in which the layers formed by SiC tetrahedrons are stacked one upon another and by the number of layers in a repeating unit of SiC layer structure. In SiC crystals synthesized in the Acheson furnace, polytypes such as 4h, 6h, 15r, or 3c are commonly observed. Their distribution varies according to the quality of the raw material used and the production conditions. The h, r, and c refer to hexagonal, rhombohedral, and cubic respectively, and the figures indicate the number of layers in a unite cell. All are polytypes of α-SiC, except for 3c, which belongs to β-SiC.

The α form is a high-temperature stable form which is generated in the high-temperature zone of 1800℃ to 2000℃ and above when synthesized in a reaction furnace. Obtained on a commercial scale as comparatively largely developed crystalline particles with a high degree of hardness, α-SiC exhibits good thermal stability, high heat resistance, and outstanding corrosion resistance. The β form is a low-temperature form generated in the low-temperature zone of 1500℃ to 1600℃. It can be obtained through a variety of method as an extremely fine powder rich in activity. The β form has a low degree of hardness and, at 1800℃ to 2000℃, it begins an irreversible transition to the α form. It is also easily oxidized. Such characteristics have made it unsuited for the applications for which SiC, in the α form, has been used in the past. Industrial production has begun, however, using specialized production facilities, for the β form as a sintering raw material powder.

Silicon carbide is synthesized commercially by the Acheson process, which involves mixing high-quality silica sand (99.5% SiO_2) with coke (carbon) in an electric resistance furnace (an Acheson furnace), with the electrodes connected by graphite powder. An electric current is passed between the poles, heating the furnace and causing the reaction to advance. The reaction within the furnace is complex but, in general, following the formula:

$$SiO_2 + 3C \longrightarrow SiC + 2CO$$

An ingot zone of α-SiC, passing through the β-SiC stage, is formed in the high-temperature zone around the graphite core. Since the further from the electrode core the lower the temperature of the zone, a cross section of the interior of the furnace after the reaction reveals an ingot zone of α-SiC with a band of β-SiC outside it; further outside there remains a layer of the unreacted raw material mixture. The lumps of α-SiC are selected out and put through the processes of crushing, decarburizing by washing, removing iron, and classification, to arrive at the product with the desired particle. To yield a fine powder with a high level of purity, a product with the appropriate grain size of green silicon carbide (GC) is used. It is normally produced by using raw material of high purity, volatilizing impurities out of them by adding salt and wood chips, and passing the resultant product through specialized fine grinding, classification, and refining processes.

The purity of the SiC can be determined based on the color of the crystals, light green 99.8%, dark green 99%, and black 98.5%.

Examples of the particle shape of both α-SiC and β-SiC fine particles are given in the SEM in Figure 2.3. The micrograph (Figure 2.3b) of β-SiC shows a preponderance of comparatively round particles, but the α-SiC particles (Figure 2.3a) are irregularly shaped, with many rather long, narrow, and angular forms, due to its being milled from coarser particles. As can be seen from the micrograph, however, depending upon the milling conditions, it is possible to produce a fine powder in a blocky shape with an elongation ratio of nearly 1.

SiC is used for high-temperature kiln furniture, electrical-resistance heating elements, grinding wheels and abrasives, wear-resistance applications, incinerator linings, varistors, light emitting diodes, and also gem stones cut from single crystals.

2.1.2.2 Silicon nitride

Silicon nitride (Si_3N_4) is another synthetic mineral. It occurs in two crystalline forms,

Figure 2.3 Scanning electron micrograph[4]
a—α-SiC; b—β-SiC

α and β, both of which belong to the hexagonal phase. The α-form is the low-temperature and the β-form is the high-temperature form. The lower temperature α-form is usually preferred as a raw material because the transformation to the β-form during sintering favors the development of an elongated crystal structure. The proportion of the α-form crystal in a powder, called the α-ratio, is a significant factor in evaluating the characteristics of these powders. A higher α ratio is preferred. Several routes are available for the synthesis of Si_3N_4 powder:

(1) Nitridation of Si powder;
(2) Carbothermal reduction of silica in N_2;
(3) Vapor phase reaction of $SiCl_4$ or silane (SiH_4) with ammonia.

Most commercially available powder is made by reacting silicon powder with nitrogen at temperatures from 1250℃ to 1400℃ according to the reaction:

$$3Si(s) + 2N_2(g) \longrightarrow Si_3N_4(s)$$

The powder generally consists of a 90 : 10 mixture of α-Si_3N_4 and β-Si_3N_4 polymorphs. Seeds of Si_3N_4 powder are often mixed with the silicon to hasten the reaction and to help prevent the formation of the undesired β phase. Nitrided powder contains impurities such as Fe, Ca, and Al originally present in the Si or picked up during subsequent milling. Higher-purity Si_3N_4 powder has been made by reduction of SiO_2 with carbon in the appropriate nitrogen environment in the range 1200 ~ 1550℃:

$$3SiO_2(s) + 6C(s) + 2N_2(g) \longrightarrow Si_3N_4(s) + 6CO(g)$$

Although this process leads to powders with residual carbon and oxygen it produces high surface area powder with a high α-Si_3N_4 content. α-Si_3N_4 seeds may again be used to speed up the reaction.

High-purity powders are also made by reaction of $SiCl_4$ or silanes with ammonia.

$$SiCl_4(g) + 6NH_3(g) \longrightarrow Si(NH)_2(s) + 4NH_4Cl(g)$$
$$Si(NH)_2(s) \longrightarrow Si_3N_4(s) + 2NH_3(g)$$
$$3SiH_4(g) + 4NH_3(g) \longrightarrow Si_3N_4(s) + 12H_2(g)$$

Powder from these reactions is amorphous, but the product on heating to 1400℃ is mostly α-Si_3N_4.

High-purity Si_3N_4 powder has also been made in small quantities by laser reaction. A mixture of gaseous silane (SiH_4) and ammonia is exposed to the coherent light output of a CO_2 laser. The silane has high absorption for the wavelengths involved, resulting in the heat required for reaction. The resulting Si_3N_4 is in spherical particles of a uniform size for the given gas flow and laser power conditions. Particles typically in the range 20 to 100 nm can be produced.

The technical indexes of Si_3N_4 powder made by Ziguang Co. (China) is list in Table 2.1, the X-ray pattern is shown in Figure 2.4. The SEM images of α-Si_3N_4 and β-Si_3N_4 powders are shown in Figure 2.5.

Table 2.1 Technical indexes of Si_3N_4 starting powder

α-Si_3N_4 content/%	Average grain size/μm	Si content/%	N content/%	O content/%	Free Si content /%
93	0.5	59	37	2	<0.8

2.1.2.3 Boron carbide

Boron carbide belongs to the important group of nonmetallic hard materials, which includes alumina, silicon carbide, and diamond. Although it was first synthesized over a century ago, in 1883, by Joly, the formula B_4C was assigned only in 1934. Today a homogeneity range from $B_{4.3}C$ to $B_{10.4}C$ has been established. The composition of commercial boron carbide is usually close to a boron : carbon stoichiometry of 4 : 1, the stoichiometric limit on the high-carbon side.

Boron carbide is typically manufactured using boric acid and graphite; after thermal conversion of boric acid to boron oxide, boron carbide is formed via:

$$2B_2O_3(l) + 7C(s) \longrightarrow B_4C(s) + 6CO(g)$$

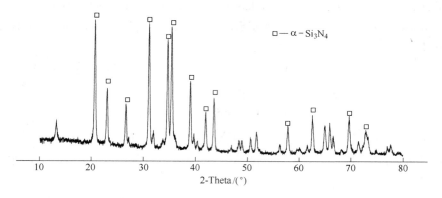

Figure 2.4 XRD pattern of Si_3N_4 powder from Ziguang Co.

Figure 2.5 SEM micrograph of Si_3N_4 powders[5]

a—α-Si_3N_4 particles; b—β-Si_3N_4 seeding particles

where l, g, s stand for liquid, gaseous and solid phases. The process is strongly endothermic, 1812kJ/mol or 9.1kW·h/kg, that is usually carried out at 1500~2500℃ in an electric furnace. Large quantities of carbon monoxide, 2.3m³/kg, are generated, and boron can be lost by evaporation of B_2O_3 at the high temperatures. An SEM image of B_4C powder made by Mudanjiang Boron Carbide Co. P. R. China is shown in Figure 2.6. The powder has a B:C ratio of about 3.9, an average particles size of 1.5μm and an oxygen content of 1.7% (mass fraction).

As an alternative fabrication method, boron carbide powders are also produced directly by magnesiothermic reductions of boric oxide in the presence of carbon at 1000~1800℃. MgO and unreacted Mg are removed by acid (H_2SO_4) washing.

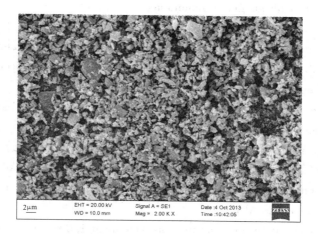

Figure 2.6 SEM image of B_4C powder

$$2B_2O_3(l) + 6Mg(s) + C(s) \longrightarrow B_4C(s) + 6MgO(s)$$

This reaction is strongly exothermic. It is carried out either directly by point ignition (thermite process) or in a carbon tube furnace under hydrogen. The product can be further purified by heating under vacuum at 1800 ℃. Since the MgO acts as a particle growth inhibitor, ultrafine boron carbide particles on the order of 0.1 ~ 1.5 μm are obtained.

B_4C has found many applications due to its extreme hardness, low density and good neutron absorption cross-section. It is used as an abrasive material in polishing and grinding media. It is also well known as an ideal control and shielding material in nuclear industry. Its extreme hardness and low density have been made it to be used in the development of lightweight armor ceramic composites. Its usage as sintered article is finding wide applications in sand blasting nozzles, ceramic bearing and wire drawing dies. In the aerospace industry, boron carbide powder is finding use as a rocket propellant due to its capacity to generate an intense amount of heat when combined with oxygen.

2.1.2.4 Titanium carbide

Titanium carbide is another nonoxide ceramic that is not available in nature. It is prepared either by the carbothermal reduction of TiO_2 or by direct reaction between the element titanium and carbon. As in many of these reactions high temperature are required. The carburization temperature is between 2100 ℃ and 2300 ℃.

2.1.2.5 Alumina nitride

There are several methods to produce AlN, two of which are currently used in industry.

One method is direct nitridation of aluminum:

$$Al(l) + 1/2 N_2(g) \longrightarrow AlN(s)$$

Al powders are converted directly to the nitride at temperatures above the melting point of the metal. Careful process control is necessary to avoid coalescence of the metal prior to nitridation.

Reducing alumina using nitrogen or ammonia in the presence of carbon is another method to produce AlN:

$$2Al_2O_3(s) + 3C(s) + N_2(g) \longrightarrow 4AlN(s) + 3CO_2(g)$$

The mixture of alumina and carbon is reacted with a nitrogen-containing atmosphere above 1400 ℃. Fine powders and extremely good control of mixing are required to result in complete conversion to AlN.

In both processes the major impurities are O (about 1.0%, mass fraction) and C (below 0.07%, mass fraction). Other impurities are silicon, iron, and calcium, which typically occur at levels below $50 \times 10^{-4}\%$ each.

Many of the applications of AlN require it to be in consolidated in the form of substrates or crucibles. It is an electrical insulator and has a high thermal conductivity (better than Fe), which makes it attractive for use in electronic packaging. Aluminum nitride crucibles are used to contain metal melts and molten salts.

2.1.2.6* Boron nitride

Although it cannot rival SiC and Si_3N_4 as a high-strength, abrasion-resistant mechanical material, hexagonal boron nitride (hBN) is a material displaying unique properties. Boron nitride can also be synthesized in a high-pressure phase (cubic phase), which is used as super-abrasive grains, as is diamond. Here, we do not discuss that phase.

An hydrous boric acid or borate (borax, for instance) is heated to 800 ℃ to 1000 ℃ with ammonia, urea, or other nitrogen compounds, making the BN bond. At this juncture, there are usually several means of controlling the reaction, such as using calcium phosphate as filler. Then the product is refined and, after further heat treatment, a thin-section BN powder with crystal growth comparable to graphitization is realized. Figure 2.7 is a scanning electron micrograph of this powder. Since the sinterability of BN is extremely poor, sintered articles are normally obtained by hot pressing, with the addition of borates, silica, or other binders.

Boron nitride is similar to graphite in structure. (See Figure 2.8). Its qualities are also similar at many points. It has quite high thermal shock resistance. It is stable up to 3000 ℃ in an inert atmosphere. It can readily be processed mechanically,

Figure 2.7 SEM image of BN powder

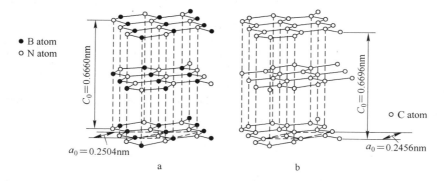

Figure 2.8 Crystal structure of BN and graphite
a—BN；b—Graphite

on a lathe for instance, and, as can be understood from its structure, it has lubricating properties.

Boron nitride also has properties not found in graphite that have led to its being valued as a unique material. In contrast to the electrical conductivity of graphite, BN has electrical insulating properties up to high temperature. In air, graphite oxidizes at 500℃ or above, but BN can be used until 900℃. Furthermore, whereas graphite reacts with metals to produce carbides, BN has outstanding resistance to corrosion, not even being wetted by most fused metals. Another distinctive feature is its white coloration.

As a powder, an oil or water suspension type and an aerogel type of BN are available for use as lubricants and as mold-release agents. Boron nitride is also used as an additive and filler in refractory, resins, and powder metallurgy.

As a sintered body, BN is employed industrially for parts and jigs for use with molten metals or glass. It is also used as a material for high-temperature furnaces, high-frequency electric insulating material, and neutron absorption materials, among others.

2.1.2.7 Zirconium diboride

ZrB_2 is useful as a crucible material for metal melts because of its excellent corrosion resistance. It is also used in Hall-Heroult cells (for Al production) as a cathode and in steel refining where it is used as thermowell tubes.

Several different processes can be used to produce ZrB_2; these are similar to those used to form carbides and nitrides. Commercially, either a direct reaction between zirconium and boron:

$$Zr + 2B \longrightarrow ZrB_2(s)$$

or carbothermal reduction of zirconia is used:

$$2ZrO_2 + C + B_4C \longrightarrow 2ZrB_2 + 2CO_2$$
$$2ZrO_2 + 5C + 2B_2O_3 \longrightarrow 2ZrB_2 + 5CO_2$$

All these reactions must be carried out at high temperature in an inert atmosphere or in vacuum.

2.1.3 Raw Material Selection

Advanced ceramic products in general are made from a high-density sintered body with a homogeneous microstructure and few holes. To yield such high-quality products, nothing is more essential that strict selection and control of the initial materials. Purity, particle size distribution, reactivity, and polymorphic form can all affect the final properties and thus must be considered from the outset.

2.1.3.1 Purity

Purity of raw materials strongly influences high-temperature properties such as strength, stress rupture life, and oxidation resistance. The effect of the impurity is dependent on the chemistry of both the matrix material and the impurity, the distribution of the impurity, and the service conditions of the component (time, temperature, stress, environment). Impurities present as inclusions do not appreciably affect properties such as creep or oxidation, but do act as flaws that can concentrate stress and decrease component tensile strength. The effect on strength is dependent on the size of the inclusion compared to the grain size of the ceramic and on the relative thermal expansion and elastic properties of the matrix and inclusion.

The effects of impurities are important for mechanical properties, but may be even more important for electrical, magnetic, and optical properties. Electrical, magnetic, and optical properties are usually carefully tailored for a specific application, often by closely controlled addition of a dopant. Slight variations in the concentration or distribution of the dopant severely alter the properties. Similarly, the presence of unwanted impurities can poison the effectiveness of the dopant and cause the device to operate improperly. For instance, in alumina ceramics, which are used in electronic devices, the presence of alkaline metallic ions is to be avoided because they lower the ceramic product's electrical insulating ability (Figure 2.9).

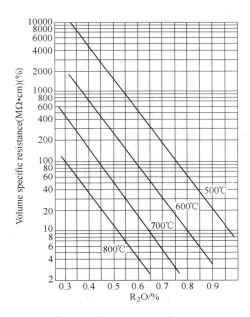

Figure 2.9 Relationship between alkaline content and insulation resistance in 90 alumina ceramics[6]

2.1.3.2 Particle size and reactivity

The particle size of starting materials has a significant effect on the properties of the properties of sintered body. In general, the finer the particle size and the larger the specific surface area, the higher the density of the sintered body that can be obtained. Conversely, the density of the green body is lower, so that shrinkage during sintering is greater. Fine powders with a specific surface area larger than necessary are, contrary to expectation, difficult to handle, and at present a specific surface area of about $15 m^2/g$ is recommended.

The research results of N. Liu and co-workers[7] shown that when starting powders are mixed with from micron particles ($2\mu m$) to sub-micron particles ($320 nm$), the microstructure of Al_2O_3-TiC composites become finer obviously. The bending strength increases from 554 MPa to 712 MPa and the fracture toughness increases also, from $5.01 MPa \cdot m^{1/2}$ to $5.36 MPa \cdot m^{1/2}$.

Particle size distribution is important, depending on which consolidation or shaping technique is to be used. In most cases the objective of the consolidation step is to achieve maximum particle packing and uniformity, so that minimum shrinkage and retained porosity will result during densification. A single particle size does not produce good packing. Optimum packing for particles all the same size results in over 30% void space. Adding particles or a size equivalent to the largest voids reduces the void content to 26%. Adding a third, still smaller particle size can reduce the pore volume to 23%. Therefore, to achieve maximum particle packing, a range of particle sizes is required.

Real ceramic particles are generally irregular in shape and do not fit into ideal packing. Porosity after compaction of these powders is generally greater than 35% and sometimes greater than 50%. Large amounts of porosity are difficult to eliminate during densification. A high degree of porosity in the compact resulted in high porosity and large grain size after sintering. Low porosity and fine grain size are beneficial to achieve a ceramic with high strength.

Reactivity is another important aspect of the starting powder. The primary driving force for densification of a compacted powder at high temperature is the change in surface free energy. Generally, as the size decreases below $1\mu m$, the particles exhibit a greater tendency to interact, and thus have a strong thermodynamic drive to decrease their surface area by bonding together. Very small particles, approximately $1\mu m$ or less, can be compacted into a porous shape and sintered at a high temperature to near-theoretical density (T.D.). An example is sintered Si_3N_4. Starting powder of approximately $2\mu m$ average particle size only sinters to about 90% of T.D. Submicron powder with a surface area roughly greater than $10 m^2/g$ sinters to greater than 95% of T.D.

Particle size distribution and reactivity are also important in determining the temperature and the time at temperature necessary to achieve sintering. Typically, the finer the powder and the greater its surface area, the lower are the temperature and the shorter the times at temperature for densification are. This can have an important effect on strength. Long times at temperature result in increased grain growth and lower strength. To optimize strength, a powder than can be densified quickly with minimal grain growth is desired.

2.1.3.3 Packing characteristics of the powder

The packing characteristics of the powder also strongly influence the properties of the end product. The pressed density is measured by placing a 15g sample in a ϕ30mm die and press forming it by the application of 1 metric t/cm^2 of pressure for 30 seconds. This value varies according to the mean particle size and the particle-size distribution. For a given particle-size distribution, the closer the particle shape is to spherical, the higher this value (note also the specific surface area). It can be stated that a powder with a high pressed density is preferable for sintering, because it is easy to achieve a dense green body with it. The density of the sintered body, however, is affected by the surface activity of the particles and by several factors that control the mechanism for diffusion of the atoms during sintering. Therefore, it is necessary to evaluate many aspects of the characteristics of the powder.

2.2 Preparation of Powders

Many methods are available for preparation of ceramic powders. These can be divided into just three basic types: mechanical, chemical and vapor phase.

Powders consist of an assemblage of small units with certain distinct physical properties. These small units, loosely referred to as particles, can have a fairly complex structure. A variety of terms have been used to describe them and this has led to some confusion in the literature. Before discussing these preparation methods of powders, we conside some terms related powders[8].

Primary particles are the smallest clearly identifiable unit in the powder. Primary particles may be crystalline or amorphous and cannot easily be broken down into smaller units. It may be defined as the smallest unit in the powder with a clearly defined surface. For a polycrystalline primary particle, the crystals have been referred to variously as crystallites, grains, or domains. In this book, we shall use the term crystallite.

Agglomerates are clusters of bonded primary particles held together by surface forces, by liquid, or by a solid bridge. Figure 2.10 is a schematic diagram of the agglomerate consisting of dense, polycrystalline primary particles. Figure 2.11 is an SEM image of a Al_2O_3 agglomerate. Agglomerates are porous, with the pores being generally interconnected. They are classified into two types: soft agglomerates and hard agglomerates. Soft agglomerates are held together by fairly weak surface forces and can be broken down into primary particles by ultrasonic agitation in a liquid. Hard agglomerates consist of primary particles that are chemically bonded by solid bridges; they therefore cannot be broken down into primary particles by ultrasonic agitation in a liquid. Hard agglomerates, are undesirable in the production of advanced ceramics because they commonly lead to the formation of microstructural defects and should be avoided in ceramic powder processing as much as possible.

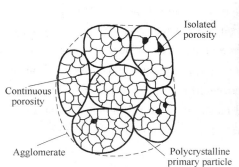

Figure 2.10 Schematic diagram of an agglomerate consisting of dense, polycrystalline primary particles

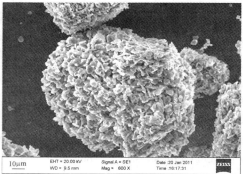

Figure 2.11 SEM image of Al_2O_3 agglomerate

Particles are a general term applied to both primary particles and agglomerates. When no distinction is made between primary particles and agglomerates, the term particles is used.

Granules are large agglomerates, usually 0.1 ~ 1mm in diameter, which are formed by the addition of a granulating agent (e.g., a polymer binder). The mixture is tumbled, producing large, nearly spherical granules that flow freely and can be used to fill complex molds and in automated processes.

Flocs are clusters of particles in a liquid suspension held together electrostatically. The formation of flocs is undesirable because it decreases the packing homogeneity of the consolidated body.

Colloids are very fine particles (they can be as small as 1nm in diameter) held in fluid suspension by Brownian motion. Consequently, colloidal particles will settle very slowly. The size range for colloidal particles is approximately 1 nm to 1 μm.

Aggregates are coarse constituents, above 1mm, in a mixture. In aggregates, a group of particles are bonded tightly by strong solid bonding formed by sintering, etc. Gentle stirring cannot break these particles into primary particles. Rigorous grinding, such as ball milling, is needed to break them. Aggregates are very detrimental for sintering.

Grain is a crystal or microcrystal in a polycrystalline material.

2.2.1 Mechanical Methods to Prepare Powders

Mechanical methods use coarse-grained materials. They are subjected to a series of processes, collectively referred to as comminution, in which the particle size is gradually reduced.

2.2.1.1 Ball milling

The most common way to achieve this size reduction is by milling. Milling produces a particular particle size distribution and deagglomeration of fine powders. Physical processes include impact, shear between two surfaces, and crushing by a normal force between two hard surfaces. When a solid is fractured, energy is given off as heat from fracture, friction in the equipment, and energy necessary to create additional surface area. It is the energy from creating additional surface area that does the work sought.

Ball mills are categorized into various types depending on the method used to impart motion to the balls (e.g. tumbling, vibration, and agitation). The advantages of ball milling are that the equipment is: simple (although experimentally straightforward, there are many theoretical aspects that are quite complex), and inexpensive. The disadvantages of ball milling are that it cannot produce ultrafine particles, can add impurities to the powder from the media and the inside of the mill, is inefficient, less than 2% of the energy input goes into creating new surfaces.

Ball milling consists of placing the particles to be ground in a closed cylindrical container (usually with ceramic or polyurethane lining) with grinding media in the form of balls, short cylinders, or rods and rotating the cylinder horizontally on its axis so that the media cascade. This is illustrated schematically in Figure 2.12. The ceramic particles move between the much larger media and between the media and the wall of the mill and are effectively broken into successively smaller particles.

For the grinding of advanced ceramics, the wall of the mill usually uses the alumina ceramics, polyurethane and various types of rubber as lining. The media should have a high density as this provides for the most effective collisions. High-specific-gravity media can accomplish a specified size reduction much more quickly than low-specific-gravity media. The ZrO_2, Al_2O_3 and SiO_2 media are commonly used. However, it is

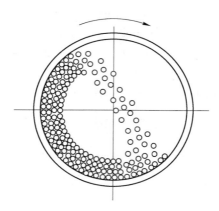

Figure 2.12 Cross section of a ball mill showing the movement of the media as the mill rotates about its axise

best that the choice of media is based on the same composition as the powder being milled.

Depending on the amount of powder to be milled, the size of the mill, and the final particle size required, the media could range from more than 8 cm in diameter to 0.6 cm, which is used for fine grinding. For a given volume, the number of balls increases inversely as the cube of the radius. Assuming that the rate of grinding depends on the number of contact points between the balls and the powder and that the number of contact points, in turn, depends on the surface area of the balls, then the rate of grinding will increase inversely as the radius of the balls. However, the balls cannot be too small since they must impart sufficient mechanical energy to the particles to cause fracture. The powder is often milled in a liquid with a surface-active agent added. Ball milling eliminates aggregates and can typically reduce the particle size down to 1 μm.

The speed of rotation of the mill is an important variable since it influences the trajectory of the balls and the mechanical energy supplied to the powder. Defining the critical speed of rotation as the speed required to just take the balls to the apex of revolution (i.e., to the top of the mill where the centrifugal force just balances the force of gravity), we find that the critical speed (in revolutions per unit time) is equal to $(g/a)^{1/2}/(2\pi)$, where a is the radius of the mill and g is the acceleration due to gravity. In practice, ball mills are operated at about 75% of the critical speed so that the balls do not reach the top of the mill (Fig. 2.12). The ball milling process does not lend itself easily to rigorous theoretical analysis. M. N. Rahaman gave an empirical relationship[8]:

2.2 Preparation of Powders

$$\text{Rate of milling} \approx A a_m^{1/2} \frac{\rho d}{r} \quad (2.1)$$

where A is numerical constant that is specific to the mill being used and the powder being milled, a_m is the radius of the mill, ρ is the density of the balls, d is the particle size of the powder, and r is the radius of the balls. According to Eq. (2.1), the rate decreases with decreasing particle size; however, this holds up to a certain point since a practical grinding limit is reached after a certain time. The variation of the rate of grinding with the radius of the balls must also be taken with caution; the balls will not possess sufficient energy to cause fracture of the particles if they are too small.

In the milling process, the objective is to have the balls fall onto the particles at the bottom of the mill rather than onto the mill liner itself. For a mill operating at about 75% of its critical speed, this occurs for dry milling for a quantity of balls filling about 50% of the mill volume and for a charge of particles filling about 25% of the mill volume. For wet milling, a useful guide is for the balls occupying about 50% of the mill volume and the slurry about 40% of the mill volume with the solids content of the slurry equal to about 25% ~ 40%. Wet ball milling has an advantage over dry milling in that its energy utilization is somewhat higher (by about 10% ~ 20%). A further advantage is the ability to produce a higher fraction of finer particles. Disadvantages of wet milling are the increased wear of the grinding media, the need for drying of the powder after milling, and contamination of the powder by the adsorbed vehicle.

Following we give some information about the preparation of 95 alumina ceramic powder. Table 2.2 lists the composition of 95 alumina ceramics which is used in practical production. All the raw materials are selected carefully.

Table 2.2 Composition of 95 alumina ceramics

Compositions	Alumina powder	Kaolin clay	Magnesia	Dolomite	Calcined talc	Barium carbonate	PVA	Dispersant	Water reducing admixture
Mass fraction/%	89	7	0.5	1	1.5	1	0.3	0.3	0.35

For the production of alumina ceramic products, milling is generally important process. Figure 2.13 is a photograph of some 15t ball mills to grind 95 alumina ceramic materials.

Following, we will introduce the milling process of alumina ceramics. Before filling the mill, every raw material is weight according to the composition of 95 alumina ceramics in Table 2.2.

Figure 2.13 Photograph of some 15t ball mills to grind 95 alumina ceramic materials

First, put PVA, dispersant, water reducing admixture and half of alumina powder into the 15t Ball Mill, to grinding for 10h. Total weight of all materials is 13.5 ~ 14 t. The ratio of raw materials : ball : water is 1 : (1.6 ~ 2.0) : (0.5 ~ 0.6). The ratio of different diameter balls of alumina is 30% (mass fraction) ϕ60mm, 40% (mass fraction) ϕ50mm, and 30% (mass fraction) ϕ40mm. For the first addition, total ball weight is about 25 t.

After grinding of 10h, the rest of materials and water are added to the ball milling to grind continuously for another 20 ~ 22h.

Before 30min of issuing slurry, measure the viscosity and finesses of the slurry. Water content: 34% ~ 35%, viscosity: >36s, D_{50}: 2.3 ~ 2.6μm, D_{90}: 4.5 ~ 5.2μm.

When all the controlling parameters can satisfy the technical requests, the slurry issues pass through a vibrating screen (as Fig. 2.14) to sieve the impurities and deironing passageway (as Figure 2.15), to a mud pool for ageing 24 ~ 72h, ready to send to spray drying.

2.2.1.2 Vibratory milling

Vibratory milling is substantially different from ball milling or attrition milling. The powder is placed in the stationary chamber of the mill together with suitable grinding media and a liquid. When the mill is turned on, vibration is transmitted (usually from the bottom center of the mill) through the chamber and into the media and powder. This results in two types of movement. First, it causes a cascading or mixing action of the contents of the milling chamber. Second, it causes local impact and shear fracturing of the powder particles between adjacent grinding media.

Figure 2.14 Photograph of a mud vibrating screen

Figure 2.15 Photograph of a deironing passageway of mud

Vibratory milling is relatively fast and efficient and yields a finer powder than is usually achieved by ball milling. The vibratory mill chamber is typically lined with polyurethane or rubber and minimizes contamination.

Figure 2.16 shows the diagrammatic view of the eccentric vibratory mill. Figure 2.17 compares the particle size reduction rate for vibratory versus rotary (ball) milling of a ferrite powder. Fine particle size was achieved rapidly by vibratory milling.

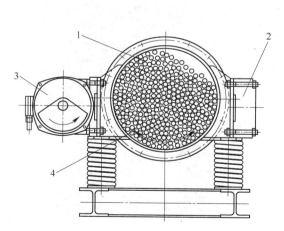

Figure 2.16 Diagrammatic view of the eccentric vibratory mill[9]
1—Grinding pipe; 2—Balancing mass;
3—Unbalanced drive; 4—Grinding balls or rods

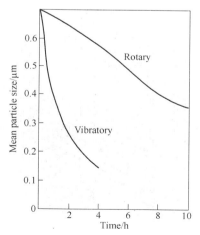

Figure 2.17 Comparison of the particle size reduction of a ferrite powder by vibratory milling and ball milling[10]

2.2.1.3 Attritors

The attritor is a grinding mill containing internally agitated media. It has been generically referred to as a "**stirred ball mill**". For efficient fine grinding, both impact action and shearing force must be present, as Figure 2.18. When wet grinding in the attritor, impact action is created by the constant impinging of the grinding media due to its irregular movement. Shearing action is present in the attritor as the balls (media) in their random movement are spinning in different rotation and, therefore, exerting shearing forces on the adjacent slurry. As a result, both liquid shearing force and media impact force are present. Such combined shearing and impact results in size reduction as well as good dispersion.

The most important concept in the attritor is that the power input is used directly for agitating the media to achieve grinding and is not used for rotating or vibrating a large, heavy tank in addition to the media.

The attritors grind faster than ball mills. Additionally, they use less of the grinding media. Most stirred mills are water jacketed to disperse the heat. Without a heat dispersal mechanism, one could overheat the binder. Figure 2.19 shows a photograph of attrition mill from Shandong Institute of Industrial Ceramics.

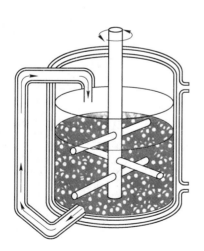

Figure 2.18 Schematic of an attrition mill[11]

Figure 2.19 Photograph of an attrition mill

Attritors have a cylindrical shell that can be lined with wear resistant materials and a rotating vertical impeller. The impeller has cross bars that can be covered with wear-resistant, ceramic sleeves. The attritor is filled about two-thirds full with the grinding media. All attritors should use a spherical media as other media shapes tend to over pack and overload the drive. The 2~3mm diameter, TZP spheres are preferred in many cases. Attritors are much more energetic than ball mills; as such, milling times are reduced. Seals and bearings are up out of the batch creating a maintenance advantage. Attritors can be operated in three modes: batch, recirculating, and continuous sometimes in series.

Table 2.3 lists the data for dry attrition milling of silicon powder. Although the attrition mill was lined with an iron-base alloy, no significant iron or carbon was picked up, even after 18 h of milling. The oxygen content did increase by interaction with air, as would normally be expected for fine particles of silicon, as the surface area increased.

Table 2.3 Changes during attrition milling of silicon powder[10]

Milling time /h	Surface area /$m^2 \cdot g^{-1}$	Iron content (mass fraction) /%	Carbon content (mass fraction) /%	Oxygen content (mass fraction) /%
0	3.0	0.62	0.03	0.60
1	11.5	0.59	0.04	2.29
4	14.5	0.58	0.04	2.57
18	23.3	0.55	0.04	3.67

Even though the average particle size was substantially reduced by dry attrition milling, some particles in the range 40 μm were still present and ultimately controlled the strength of the final component. Evidently, some particles were trapped or packed in regions of the mill where they did not receive adequate milling. This did not occur for wet milling.

A different configuration of attrition mill that has been used primarily for wet milling is referred to as a turbomill and consists of a rotating cylindrical cage of vertical bars surrounded by a stationary cylindrical cage of vertical bars. The material to be ground is mixed with water or other fluid plus sand-size grinding media. Materials such as ZrO_2, Al_2O_3, and SiO_2 have been ground to submicron size in a few hours, compared to 30 h for vibratory milling and much longer times for ball milling. The primary problem with this attrition milling approach is the amount of contamination and the difficulty of separating the powder from the media. For example, in one case 20% to 30% of the media was ground to 325 mesh (<44μm) and was not successfully separated from the 0.1 μm milled powder.

2.2.1.4 Bead milling

The bead mill consists of a grinding chamber, which in turn houses the shaft and agitator discs. The chamber is filled with zirconia grinding media and the material leakage is prevented by the double acting mechanical seal. The material is pumped into the chamber using an air operated diaphragm pump or screw pump. Figure 2.20 shows a photograph of a bead mill and Figure 2.21 shows a schematic diagram of a micro-bead type bead mill.

Figure 2.20　Photograph of a bead mill

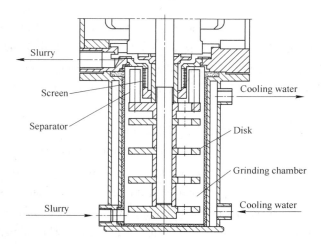

Figure 2.21　Schematic of a micro-bead type bead mill[12]

During the work procedure, the motor of equipment drives the inside feeding device rotating at high speed, which creates negative pressure. Through the self-absorption effect, materials on the top of material tank will be inhaled and delivered to the grinding chamber. The grinding media inside the grinding chamber accelerated by the high speed rotation rod run irregularly in all directions and put the particles of materials under the action of constant collision and friction between the grinding media. At the same time, the screen separates the materials and grinding media. The materials keep flowing from the container to the grinding chamber and become smooth circulation to achieve smaller particle size and narrower particle size range.

Bead mill is characterized in high grinding efficiency, good grinding fineness, wide application and good wear resistance. Bead mills look and operate like sand mills. The only difference between the two is the type of grinding media employed. While conventional sand mills ordinarily use Ottawa sand, bead mills use a wide variety of synthetic media including glass, ceramic, and zirconium oxide or zirconium silicate beads. The term "beadmilling" developed in the 1960s when manufacturers started using synthetic grinding media rather than sand. Many former "sand" mills are now "bead" mills.

Dongying Guoci Co. (China) uses bead mill to grind the powder of barium titanate, with different grain size of zirconia grinding media, the ratio of 0.45 mm to 0.65 mm of zirconia balls is 3 : 2, after grinding of 4h, the particle size of barium titanate powders changes 80 ~ 100 nm from 10 ~ 20μm. The bead milling is an effective mechanical method to prepare nano-powder.

2.2.1.5 Planetary mills

The planetary ball mill (Figure 2.22) owes its name to the planet-like movement of its vials. Since the vials and the supporting disc rotate in opposite directions, the centrifugal forces alternatively act in like and opposite directions. This causes the milling balls to run down the inside wall of the vial—the friction effect, followed by the material being milled and milling balls lifting by of and traveling freely through the inner chamber of the vial and colliding against the opposite inside wall[13].

Planetary mills exploit the principle of centrifugal acceleration instead of gravitational acceleration. The enhancement of the forces acting on the balls in relation to the conventional ball mill is achieved by the combined action of two centrifugal fields. The charge inside vials performs two relative motions: a rotary motion around the mill axis and a planetary motion around the vial axis (Figure 2.23).

 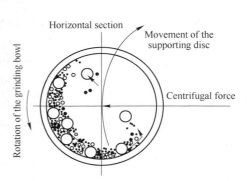

Figure 2.22 Laboratory planetary mill

Figure 2.23 Movements of working parts and balls in a planetary mill

During the past decades the problem of continuous feeding of a material into a planetary mill has been successfully solved and novel planetary mills of the industrial scale operating in a continuous mode are now manufactured. In this mode the initial material is continuously fed into the mill, with coarser powder returned for another cycle of milling and the final fine or nanoscale powder product continuously provided as a result of the milling process.

2.2.1.6 Fluid energy milling

Fluid energy milling, also called **jet milling**, achieves particle size reduction by particle-particle impact in a high-velocity fluid. The fluid can be compressed air, nitrogen, carbon dioxide, superheated steam, water, or any other gas or liquid compatible with the specific equipment design. The powder is added to the fluid and accelerated to sonic or near-sonic velocity through jets leading into the grinding chamber. The grinding chamber is designed to maximize particle-particle impact and minimize particle-wall impact.

Fluid energy milling can achieve controlled particle sizing with narrow size distribution. Most jet mills have no moving parts and can be easily lined with polyurethane, rubber, wear-resistant steel, and even ceramics. Figure 2.24 shows photograph (a) and schematic (b) of a jet mill.

Typical grinding data for fluid energy milling are shown in Table 2.4. Output capacity can range from a few-grams per hour to thousands of kilograms per hour depending on the size of the equipment.

2.2 Preparation of Powders

The main drawback with fluid energy milling is collecting the powder. Large volumes of gases must be handled. Cyclones are not efficient for micrometer-sized particles and tilters clog rapidly.

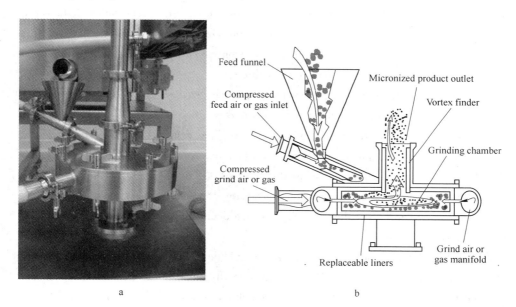

Figure 2.24 Photograph (a) and schematic (b) of a jet mill
(From Homepage of Utek International Corporation Limited)

Table 2.4 Typical grinding data for fluid energy milling[10]

Material	Mill diameter		Grinding medium	Material feed rate		Average particle size obtained	
	cm	in		kg/h	lb/h	μm	in
Al_2O_3	20.3	8	Air	6.8	15	3	0.00012
TiO_2	76.2	30	Steam	1020	2250	<1	<0.00004
TiO_2	106.7	42	Steam	1820	4000	<1	<0.00004
MgO	20.3	8	Air	6.8	15	5	0.0002
Coal	50.8	20	Air	450	1000	5~6	about 0.00025
Cryolite	76.2	30	Steam	450	1000	3	0.00012
DDT 50%	61.0	24	Air	820	1800	2~3	about 0.0001
Dolomite	91.4	36	Steam	1090	2400	<44	<0.0018
Sulfur	61.0	24	Air	590	1300	3~4	about 0.00014
Fe_2O_3	76.2	30	Steam	450	1000	2~3	about 0.0001

2.2.1.7 Powder separation

A Screening

Screening is a sorting method of particle sizing. The powder is poured onto a single screen having selected size openings or on a series of screens, each subsequently with smaller openings. The particles are separated into size range; the particles larger than the screen openings remain on the screen and smaller particles pass through until they reach a screen with holes too small to pass through.

Screening can be conducted dry or with the particles suspended in slurry. Dry screening is used most frequently for larger particles and is a fast and effective approach. It is used in the mining industry and in many phases of the ceramic industry, especially in the sizing of abrasives. For free flowing particles, dry screening can normally be effective down to about 325 meshes. Below this the particles are so fine that they either tend to agglomerate or clog the screen. Some automatic screen systems use airflow or vibration to aid in screening powders that have a significant portion of particles in the range of 325 mesh or smaller.

Suspending the particles in a dilute water or other liquid suspension (slurry) also aids in screening fine particles. Slurries can normally be screened easily through at least 500 meshes as long as the solid content in the slurry is low and fluidity is high. For very fine powder, this is a useful method of assuring that no particles larger than an acceptable limit (determined by the screen size selected) are left in the powder. Since isolated large particles in a powder of fine particle size distribution often become the strength-limiting flaw in the final component, wet screening can be used as an in-process quality control step.

Screen sizes are classified according to the number of openings per linear inch and are referred to as mesh sizes. A 16-mesh screen has 16 equally spaced openings per linear inch; a 325-mesh screen has 325. Table 2.5 compares the mesh size of standard screens with the actual size of the openings.

Table 2.5 ASTM standard screen size

"Mesh" seive designation	4	6	10	12	16	20	40	80
Seive opening/mm	4.76	3.36	2.00	1.68	1.19	0.84	0.42	0.177
"Mesh" seive designation	120	170	200	230	270	325	400	
Seive opening/mm	0.125	0.088	0.074	0.063	0.053	0.044	0.037	

Screening does have limitations. If the powder tends to compact or agglomerate, groups of particles will act as a single particle and result in inaccurate screening. Similarly, packing or agglomeration can clog the screen and prevent further screening or decrease efficiency.

B Air classification

Air classification (also referred to as air separation) is used to separate coarse and fine fractions of dry ceramic powders. A schematic of an air classifier is shown in Figure 2.25.

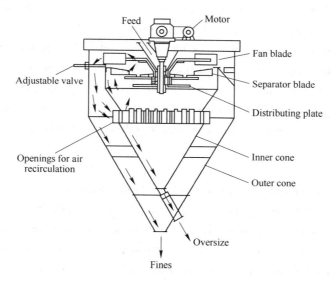

Figure 2.25 Drawing of an air classifier, showing the paths of the over size and fine particles

Separation is achieved by control of horizontal centrifugal force and vertical air currents within the classifier. Particles enter the equipment along the centerline and are centrifugally accelerated outward. As the coarse particles move radially away from the center into the separating zone, they lose velocity and settle into a collection cone. The finer particles are carried upward and radially by the air currents through selector blades. These selector blades impart an additional centrifugal force to the particles and cause additional coarse particles to settle into the coarse collection cone. The fines are then carried by the airflow to a separate cone for collection.

Air classification is frequently linked directly to milling, crushing, grinding, or other comminution equipment in a closed circuit. Particles from the mill are discharged directly into the air classifier. The fines are separated and the coarse is returned to the mill for further grinding. One type of unit combines size reduction and classification into a single piece of equipment. The coarse powder particles are carried by high-velocity air through two opposing nozzles. Where the two air streams meet particles strike each other and are shattered into smaller particles. The air carrying the particles flows vertically. Large particles pass through a centrifugal-type air classifier at the top of the unit, where additional controlled sizing is accomplished.

Air classification has its advantages and limitations. It is an efficient and high-volume approach for separating coarse particles from fine particles and producing controlled size ranges roughly from 40 to 400 mesh. However, it is limited in its efficiency and accuracy in producing controlled sizing of particles below 10μm.

In air classification, one problem is the presence of contamination. Sliding motion and impact of the ceramic particles against equipment surfaces result in some metallic contamination. The amount of contamination is less than for comminution equipment, but may still cause a problem for some applications. Therefore, we should be aware of the air classification process and its potential for affecting material application specification.

C   Elutriation

Elutriation is a general term that refers to particle size separation based on settling rate; that is, large or high-specific-gravity particles settle more rapidly from a suspension than do small or low specific-gravity particles. Air classification is a form of elutriation where the suspending medium is air. For this book the term elutriation is used to describe particle sizing by settling from a liquid suspension.

Elutriation is frequently used in the laboratory for obtaining very fine particle distributions free of large particles. The powder is mixed with water or other liquid and usually with a wetting agent and possibly a deflocculant to yield a dilute suspension. Stirring or mixing is stopped and settling is allowed to occur for a predetermined time. The time is based on the particle size cut desired. The fluid containing the fine particles is then decanted or siphoned and the remaining fluid and residue discarded or used for some other purpose.

A major problem with elutriation is that the fines must be extracted from the fluid before they can be used. This can be done by evaporating the fluid or by filtration. Both tend to leave the fines compacted or crusted rather than as a free-flowing powder, thus requiring additional process steps before the powder can be used. Also, unless the elutriation and liquid removal are conducted in a closed system, chances for contamination are high.

2.2.2 Powder Synthesis by Chemical Methods

The synthesis of new ceramic materials and the development of novel fabrication techniques are expected to be key factors for further optimization of properties, especially at high temperatures. Besides the conventional powder processing route, the chemical formation of high-purity advanced ceramics has been utilized by chemical vapour deposition (CVD), sol-gel technology and polymer pyrolysis. In the case of CVD, high-volatile compounds are decomposed in the gas phase and precipitated as oxide or non-oxide ceramics, as well as metals or semi-conductors, depending on the starting material. In contrast, the sol-gel and the polymer pyrolysis process employs non-volatile inorganic-organic polymers to generate both oxide and non-oxide ceramics, preferentially via the liquid or solid state[14~16].

2.2.2.1 Mechanochemical synthesis

Powder preparation by high-energy ball milling of elemental mixtures is referred to by various terms, including **mechanochemical synthesis, mechanosynthesis, mechanical driven synthesis, mechanical alloying,** and **high energy milling**[8]. While no term has received widespread acceptance, the term **mechanochemical synthesis** is adopted in this book. Since this process combines the mechanical energy with chemical action, here we categorize it to chemical method.

Mechanochemical processing route has attracted significant interest in the last 20 years or so for the production of intermetallic and alloy compounds and some ceramic materials[17].

The mechanical activation is one of the most effective method for obtaining highly disperse system due to mechanical action stress fields form in solids during milling procedure. This effect results in changes of free energy, leading to release of heat, formation of a new surface, formation of different crystal lattice defects and initiation of solid-state chemical reaction. The accumulated deformation energy determines irreversible changes of crystal structure and consequently microstructure resulting in the change of their properties. The intrinsic advantage of this process is that the solid-state reaction is activated due to mechanical energy instead of the temperature. It was shown that the chemical reactivity of starting materials could be improved significantly after mechanochemical activation and, subsequently, the calcination temperature was reduced. Besides, it was apparent that the mechanochemical treatment could enhance the reactivity of constituent oxides[18].

An advantage of mechanochemical synthesis is the ease of preparation of powders that can otherwise be difficult to produce, such as those of the silicides and carbides. For example, most metal carbides are formed by the reaction between metals or metal hydrides and carbon at high temperatures (in some cases as high as 2000℃). Furthermore, some carbides and silicides have a narrow compositional range that is difficult to produce by other methods. A disadvantage is the incorporation of impurities from the mill and milling medium into the powder.

The mechanism of mechanochemical synthesis is not clear. One possibility is the occurrence of the reaction by a solid-state diffusion mechanism. Since diffusion is thermally activated, this would require a significant lowering of the activation energy, a considerable increase in the temperature existing in the mill, or some combination of the two. While considerable heating of the mill occurs, the temperature is significantly lower than that required for a true solid-state mechanism. A second possibility is that the reaction occurs by local melting during the milling process. While melting of the particles may accompany highly exothermic reactions, as outlined for the next mechanism, the evidence for compound formation by local melting is unclear. A third possibility is the occurrence of the reaction by a form of selfpropagating process at high temperature. In highly exothermic reactions, such as the formation of molybdenum and titanium silicides from their elemental mixtures, the heat that is liberated is often sufficient to sustain the reaction. However, for the reaction to first occur, a source of energy must be available to raise the adiabatic temperature of the system to that required for it to become self-sustaining. The surface energy of the very fine powders prior to extensive reaction is quite enormous[8].

Even though the mechanisms of the energy exchange during these processes are not completely clear, they have efficient contribution to mechanical activation, generally leading to physical and in some cases to chemical changes in material[19].

Lee and co-workers[20] used the mechanochemical technique for synthesizing fine $PbZrTiO_3$ powders. The process started from oxide powders and the perovskite phase of PZT was formed at room temperature in a mechanochemical cell, rather than via calcination at an elevated temperature. The PZT composition selected for that study was Pb($Zr_{0.52}Ti_{0.48}$)O_3 which is near the morphotropic phase boundary. The mixture of powders was grinding in ball mill for 48 h, and after drying, and sieving, mechanical synthesis was conducted in shaker-mill for 5, 10, 15 and 20 h. After activation of 5 h almost all the sharp peaks of PbO, ZrO_2 and TiO_2 had been vanished and replaced by a few broadening peaks implied that there was a rapid reduction in the particle size of constituent oxides and some degree of amorphization occurred. This was directly reflected by a dramatic increase in the specific surface area of the powder. The perovskite PZT phase formation can be considered as a consequence of nucleation in the nanocrystalline/amorphous matrix of mixed oxides, followed by crystalline growth, as a result of prolonged mechanical activation.

2.2.2.2 Solid state reactions

Chemical reactions between solid starting materials, usually in the form of mixed powders, are common for the production of powders of complex oxides such as titanates, ferrites, and silicates. The reactants normally consist of simple oxides, carbonates, nitrates, sulfates, oxalates, or acetates. An example is the reaction between zinc oxide and alumina to produce zinc aluminate:

$$ZnO(s) + Al_2O_3(s) \longrightarrow ZnAl_2O_4(s)$$

These methods, involving decomposition of solids or chemical reaction between solids are referred to in the ceramic literature as calcination.

We synthesized $Y_{6-x}Nd_xMoO_{12+\sigma}$ pigments with high near-infrared reflectance via solid state reaction route[21]. The starting materials chosen for the preparation of neodymium doped Y_6MoO_{12} composite pigments are chemically pure Nd_2O_3 (99.9%), Y_2O_3 (99.9%), and MoO_3 (99.9%). Various amount of Nd_2O_3 and Y_2O_3 were added to the MoO_3 powders according to the formula $Y_{6-x}Nd_xMoO_{12+\sigma}$ ($x=0, 0.2, 0.4, 0.6, 0.8, 1$). The mixtures were transferred to an agate mortar and mixed thoroughly in ethanol. After evaporating ethanol, the mixed powders were calcined in alumina crucibles in a muffle furnace at 1500℃ for 6h with a heating rate of 5℃/min in air atmosphere, and then auto-cooled in the furnace. Figure 2.26 shows the NIR reflectance spectra of $Y_{6-x}Nd_xMoO_{12+\sigma}$ (x ranges from 0 to 1) powder pigments calcined at 1500℃. Figure 2.27 shows the SEM image of $Y_{5.4}Nd_{0.6}MoO_{12+\sigma}$ sample calcined at 1500℃.

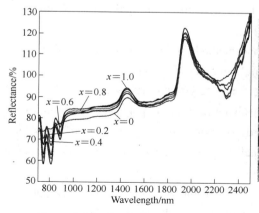
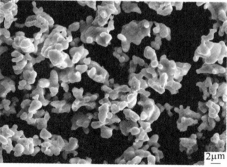

Figure 2.26 NIR reflectance spectra of $Y_{6-x}Nd_xMoO_{12+\sigma}$ (x ranges from 0 to 1) powder pigments

Figure 2.27 SEM image of $Y_{5.4}Nd_{0.6}MoO_{12+\sigma}$ sample calcined at 1500℃

Boron carbide can be prepared by reaction of elemental boron and carbon powder via carbothermal reduction process. Boron carbide is typically manufactured using boric acid and graphite; after thermal conversion of boric acid to boron oxide, boron carbide is formed via:

$$4H_3BO_3 + 7C = B_4C + 6CO(g) + 6H_2O(g)$$

The process is strongly endothermic, 1812kJ/mol or 9.1kW · h/kg, that is usually carried out at 1500~2500℃ in an electric furnace. Either arc furnaces or resistance furnaces, analogous to Acheson furnaces for SiC, are used. The starting material is an intimate mixture of boric oxide and carbon, e.g., petroleum coke or graphite. Large quantities of carbon monoxide, $2.3m^3/kg$, are generated, and boron can be lost by evaporation of B_2O_3 at the high temperatures.

Silicon nitride powder is also produced by the carbothermic reduction of SiO_2 in a mixture of fine SiO_2 and C powders followed by nitridation between 1200℃ and 1400℃ in N_2. This process is used industrially by Toshiba (Japan). The widespread availability of pure, fine SiO_2 and C makes this method an attractive alternative to the nitridation of Si. While the overall reaction can be written: $3SiO_2 + 6C + 2N_2 \longrightarrow Si_3N_4 + 6CO$, the mechanism is believed to involve the gaseous silicon monoxide, SiO, as follows[8]:

$$3SiO_2(s) + 3C(s) \longrightarrow 3SiO(s) + 3CO(g)$$
$$3SiO(s) \longrightarrow 3SiO(g)$$
$$3SiO(g) + 3C(s) + 2N_2 \longrightarrow Si_3N_4(s) + 3CO(g)$$

Excess carbon is used as an oxygen sink to form gaseous CO and reduce the amount of oxygen on the powder surface. However, any C remaining after the reaction has to be burnt out in an oxidizing atmosphere, and this may cause some reoxidation of the Si_3N_4 surfaces.

The nitridation and carbothermic reduction methods produce a strongly agglomerated mass of Si_3N_4 that requires milling, washing, and classification.

2.2.2.3 Synthesis of ceramic powders by polymer pyrolysis

The polymer pyrolysis of appropriate organosilicon preceramic polymers has been known as a practicable method to produce Si-based non-oxide ceramics at noticeably low temperatures (80~1500℃)[22]. This chemical route of ceramic fabrication is composed of:

(1) The synthesis of oligomers or polymers from low molecular compounds (precursors), which consist of structural elements as desired in the final product.

(2) Chemical or thermal cross-linking of the as-synthesized precursors in order to obtain high molecular compounds convertible into ceramics with high yields.

(3) Pyrolysis of the cross-linked polymer providing the desired ceramic material, accompanied by the formation of gaseous reaction products.

The main advantages of this production route are the easy purification of starting materials and thus the extraordinarily effective reduction of impurities in the final product, the adjustment of the viscosity of the polymers in a wide range and the generation of novel, metastable, amorphous materials that cannot be produced by conventional methods, as well as the low fabrication temperatures. This enables not only the manufacture of extremly creep-and oxidation-resistant ceramic materials for high-temperature applications, but also the processing of the polymers by well-established industrial methods such as coating of substrates and spinning of fibres.

The author and his co-workers synthesized Er-dopes SiC/SiO_2 nano-composites via polymer pyrolysis and investigated the optical characteristics[23]. Triethoxysilane ($HSi(OCH_2CH_3)_3$, T^H) and methyldiethoxysilane ($HCH_3Si(OCH_2CH_3)_2$, D^H), and $ErCl_3 \cdot 6H_2O$ were as raw materials, THF as solvent. The molar ratio of $T^H/D^H = 2$ chosen was to lead, after pyrolysis, to the formation of a pure silicon oxycarbide glass with negligible amount of free C. In a typical preparation, the alkoxides were mixed in THF using an ice bath. Then the acidic water was added drop by drop under vigorous stirring. Finally, the $ErCl_3$ solution was added in order to achieve the desired Er content ($Er/Si = 1\%$) and to adjust a total $H_2O/O(CH_2CH_3) = 1$ ratio. This solution was stirred for additional 20 min and then cast into small plastic tubes and sealed with plastic film. Gelation occurred within 2h at room temperature. Gels were then slowly dried at room temperature for 3 weeks and then at 80℃ for an additional week. Monolithic, cylindrical samples 6mm × 30mm were accordingly obtained. From these samples thin gel discs, 0.5~1mm thick were subsequently obtained by sectioning the original monolithic gels. Pyrolysis was performed in an alumina tubular furnace in flowing Ar (100mL/min) at 800, 1000, 1200 and 1300℃. The results indicated that erbium activated SiC/SiO_2 nanocomposites prepared by polymer pyrolysis has an emission in the C-telecommunication band at room temperature upon continuous-wave excitation at 980 nm. The spectral bandwidth of 48 nm is large enough for application in wavelength division multiplexed (WDM) signal amplifiers.

2.2.2.4 Synthesis of ceramic powders by sol-gel routes

The term **sol-gel** is used broadly to describe the preparation of ceramic materials by a process that involves the preparation of a sol, the gelation of the sol, and the removal of the liquid. The first sol-gel synthesis of silica was described by Ebekmen in 1844 and the commercial production of sol-gel coatings onto flat glass appeared in the sixties.

However, the development of sol-gel science only really started with the "First International Workshop on Glass and Ceramics from Gels" in 1981[24]. Sol-gel processes now widely used for synthesis of multi component ceramics, fiber, film, nano-wire, nano-tubes and many papers have been published every year.

In the sol-gel process, a **sol** is a suspension of colloidal particles in a liquid or a solution of polymer molecules. The term **gel** refers to the semirigid mass formed when the colloidal particles are linked to form a network or when the polymer molecules are cross-linked or interlinked. Two different sol-gel processing routes are commonly distinguished: the particulate (or colloidal) gel route in which the sol consists of dense colloidal particles (1 to 1000 nm) and the polymeric gel route in which the sol consists of polymer chains but has no dense particles about 1nm. In many cases, particularly when the particle size approaches the lower limit of the colloidal size range, the distinction between a particulate and a polymeric system may not be very clear[8].

The starting compounds (precursors) for the preparation of the sol consist of inorganic salts or metal-organic compounds, but the metal alkoxy compounds are to be used in most sol-gel processes. Alkoxides result from the reaction of metals (Me) with alcohols. The general reaction is:

$$n\text{ROH} + \text{Me} \longrightarrow (\text{RO})_n\text{Me} + (n/2)\text{H}_2$$

where R is an organic group. For ethanol, R is the ethoxy group C_2H_5. Catalysts are often necessary to increase reaction rates. For example, aluminum will react with isopropanol at 80℃ in the presence of a small amount of $HgCl_2$. In this case the catalyst breaks down the protective oxide layer that forms on the aluminum. A number of metal alkoxides are commercially available in high purity form. We can buy the metal alkoxides to prepare the oxide powders.

Preparing a powder by a sol-gel approach involves following steps:

(1) Form a stable dispersion (sol) of particles less than 0.1 μm in diameter in a liquid.

(2) By change in concentration (evaporation of a portion of the liquid), aging, or addition of a suitable electrolyte, induce polymer-like, three-dimensional bonding to occur through out the sol to form a gel.

(3) Evaporate the remaining liquid from the gel.

(4) Increase the temperature to convert the dehydrated gel to the ceramic composition.

Following we use aluminum isopropoxide as an example to illuminate the synthesis of alumina powder by sol-gel processing. To make metal oxide powders from these organometallic precursors we start with an aluminum isopropoxide solution in alcohol. Water is added to the alcohol solution. Two reactions then occur:

$$(C_3H_2O)_2\text{-Al-OC}_3H_2 + \text{HOH} \longrightarrow (C_3H_2O)_2\text{-Al-OH} + C_3H_2OH$$
$$(C_3H_2O)_2\text{-Al-OH} + C_3H_2O\text{-Al-}(C_3H_2O)_2 \longrightarrow (C_3H_2O)_2\text{-Al-O-Me-}(C_3H_2O)_2 + C_3H_2OH$$

The remaining alkoxy groups (—OR) of the condensation product can be hydrolyzed further to form a crosslinked, three-dimensional network of metal-oxygen bonds. The actual reactions that occur appear to be significantly more complex than this two represented above.

There are several variables in the sol-gel process[25]:

(1) Rates of hydrolysis and condensation (relative differences in the rates can be used to modify the microstructure of the powder).

(2) Type of alkoxide (mixing of the alkoxides in the solution is achieved at a molecular level giving the powders a high degree of chemical homogeneity).

(3) Reaction temperature (affects the degree of polymerization of the gel).

(4) Amount of water added (affects the degree of polymerization of the gel).

(5) Solution pH (rates of hydrolysis and condensation can be increased by the addition of acids or bases respectively).

Gelation times vary from seconds to several days. When the gel forms it may contain only about 5 vol% of the oxide. The dried gel is calcined to completely convert it to oxide. Powders produced by the sol-gel method are amorphous. A crystallization step is required to produce crystalline bodies, which is often performed after sintering.

Since the rigidity of the gel prevents migration or segregation of atoms during drying, make the homogeneity of the powder composition keep at the molecular level. The resulting powder has high surface area and small particle size. Particle size is generally in the range of 20 to 50 nm. Surface areas as high as 500 m^2/g have been reported. Due to the high surface area and fine particle size, these sol-gel-derived powders can be densified at lower temperatures than powder prepared by conventional mechanical processes.

The author synthesized the Eu^{2+}-doped Silicon oxycarbide glasses and investigated the luminescent properties via sol-gel technique[26]. Recently we have also synthesized $Dy_2Ti_2O_7$ nano powders by sol-gel method[27]. A powder of nano-dysprosium titanate was synthesized using tetrabutyl titanate, dysprosium nitrate as the raw materials via sol-gel process. After drying the wet gel at 50~80℃, the dried gel was heat treated at 800~1100℃, the dysprosium titanate nanopowder can be obtained. The research results show that the dried $Dy_2Ti_2O_7$ gel is highly crystallized after heat treatment at 1000℃ with an average grain size of 100 nm as Figure 2.28. Except for its use as an absorber material for control rods, we proved that it was a promising sintering aid for the densification of B_4C ceramics and improving the neutron absorption properties of B_4C neutron absorber according to the experiment result of our reseach group[28].

We have also synthesized a dysprosium aluminum garnet (DAG) nanopowder by aqueous sol-gel method, using Al powder, HCl and $Dy(CH_3COO)_3 \cdot 4H_2O$ as raw materials. The gel calcined from 900 to 1200℃ resulted in the formation of a crystalline DAG nanopowder with particle size distribution ranges from 30 to 100 nm as Figure 2.29[29].

Figure 2.28 SEM micrograph of $Dy_2Ti_2O_7$ powders heat treated at 1000℃

Figure 2.29 SEM micrograph of DAG powders heated at 1100℃

2.2.2.5 Precipitation

Precipitation of soluble salts followed by thermal decomposition to the oxide is a widely used method of both particle sizing and purifying of oxide ceramics. Analogous techniques in controlled atmospheres have also been used to produce nonoxide ceramic powders.

To cause precipitation it is necessary to produce a supersaturated solution. This can be achieved, for example, by changing the pH or the temperature. A larger quantity of a soluble component (for example, a metal salt) can be dissolved in a solution at high temperature than at a lower temperature. At a supersaturation that exceeds the concentration threshold for homogeneous nucleation, a large number of nuclei form suddenly. Their formation lowers the solution concentration below the concentration at which nucleation occurs, but enough excess solute remains for the existing nuclei to grow. If the solution is kept uniform, growth of all the particles proceeds at the same rate, producing powders with extremely uniform size distribution.

Precipitation of mixed oxides is possible. For example, in the fabrication of nickel ferrite (a magnetic ceramic used for memories) a mixed aqueous solution of iron and nickel sulfates is used. The solution is kept at about 80℃ and precipitation occurs when the pH is increased to around 11 with ammonium hydroxide. A mixed hydroxide precipitates, which is washed to remove the residual sulfate and dried to a powder with a particle size between 50 nm and 1 μm.

The Pechini method refers to an original process developed by Pechini for the preparation of titanates and niobates for the capacitor industry. The method has since been applied to many complex oxide compositions. Figure 2.30 shows a flowchart for the preparation of strontium titanate powder. Metal ions from starting materials such as carbonates, nitrates, and alkoxides are complexed in an aqueous solution with carboxylic acids such as citric acid. When heated with a polyhydroxy alcohol, such as ethylene glycol, polyesterification occurs, and on removal of the excess liquid, a transparent resin is formed. The resin is then heated to decompose the organic constituents, ground, and calcined to produce the powder.

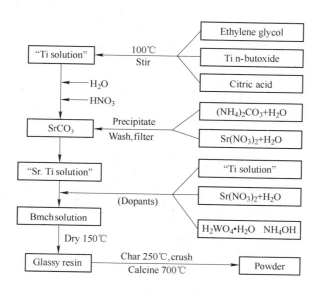

Figure 2.30 Flowchart for preparing $SrTiO_3$ powders by the Pechini method[25]

2.2.2.6 Hydrothermal technology

The term hydrothermal processing usually refers to any heterogeneous reaction in the presence of aqueous solvents or mineralizers under high pressure and temperature conditions to dissolve and recrystallize (recover) materials that are relatively insoluble under ordinary conditions. Typical temperatures range from 100 to 350℃ at pressures up to 15 MPa. Under these conditions, a wide variety of pure, fine-particle ceramic compositions can be synthesized. The feedstock can be oxides, hydroxides, salts, gels, organics, acids, and bases. The conditions can be oxidizing or reducing. The particle size can be controlled by residence time, temperature, and pressure. The resulting powder consists of single crystals of the final composition. No heat treatments or milling operations are required[30].

K. Namratha and K. Byrappa[31] introduced recently their hydrothermal experiments to prepare TiO_2 and ZnO nanoparticles and a flowchart of the experimental methodology, as Figure 2.31. The experiments were carried out in General Purpose autoclaves using

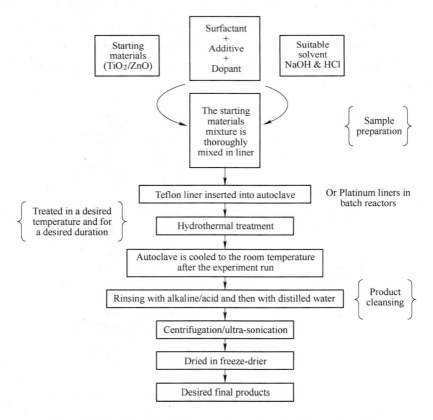

Figure 2.31 Flowchart of the experimental methodology adopted in synthesis of TiO_2 and ZnO nanopaticles[31]

Teflon liners (for sub-critical hydrothermal experiments in the temperature range 150 ~ 240℃) or Batch reactors (for super-critical hydrothermal experiments at temperature 400℃) provided with platinum or gold liners or capsules. A wide range of solvents and surfactant have been used for both TiO_2 and ZnO nanoparticles fabrication. Also the molar concentration of the raw material TiO_2 and ZnO was varied to investigate the quality of the resultant products. The raw materials were reagent grade TiO_2 and ZnO. The experimental duration was varied from 24h to just 2h depending upon the precursors used, especially based on the organic additives and surfactants. The dopants were introduced in the solution form in the desired molar concentration, and in some cases they were introduced as metals directly into the hydrothermal and solvother-mal systems. After the experimental run, the products were washed and freeze-dried.

The hydrothermal processing of advanced materials is a highly interdisciplinary subject and the technique is popularly used by physicists, chemists, ceramists, hydrometallurgists, materials scientists, engineers, biologists, geologists, technologists, and so on. Hydrothermal technology in the 21st century is not just confined to the crystal growth or leaching of metals, but it is going to take a very broad shape covering several interdisciplinary branches of science.

K. Byrappa and T. Adschiri drew a conclusion in their paper[32]: Advanced material processing using hydrothermal technology has lots of advantages owing to the adaptability of the technique, which is also environmentally benign. The use of non-aqueous and a host of other mixed solvents employed in materials processing have brought down the Pressure temperature (PT) conditions of advanced materials processing. The use of thermodynamic computation helps to intelligently engineer the materials processing and makes it the most cost effective technique even for high hardness or superhard and ultralow solubility materials. The great advantages of hydrothermal technology for nanomaterials processing are the production of particles that are monodispersed with total control over their shape and size in addition to their chemical homogeneity with the highest dispersibility. A great variety of advanced nanomaterials whether nanoparticles, or nanocomposites covering metals, metal oxides, semiconductors, silicates, sulphides, hydroxides, tungstates, titanates, carbon, zeolites, ceramics, composites, etc have been processed using hydrothermal technology. The use of multi-energy systems like microwaveehydrothermal, or electrochemicalehydrothermal, or mechanochemicalehydrothermal drives this technology to a new and totally unexplored avenue in the 21st century. Also the use of capping agents, surfactants and other organic molecules contribute greatly to the surface modification of these nanocrystals to obtain the desired physico-chemical characteristics. In fact the lowering of PT conditions of nanomaterials processing has made it the most environmentally friendly and effective technique enabling maximum yields. Hydrothermal technology bears a special mention in advanced materials processing through its ability to significantly accelerate the kinetics of synthesis, to model the theoretical approaches from solutions, to develop in situ observation techniques, to evolve supercritical water (SCW) and supercritical fluids (SCF) technologies for decomposing and recycling toxic hazardous chemical wastes as well as the monomerization of high polymer wastes and other environmental engineering and chemical engineering issues like recycling of rubbers and plastics instead of burning, etc. The combination of hydrothermal technology and nanotechnology can answer most of the problems associated with advanced materials processing in the 21st century.

2.2.2.7 Synthesis of ceramic powders by gel-casting route

Recently, the gel-casting technique, which is a novel forming method in fabricating complex three-dimensional ceramic parts has been employed to synthesize ceramic powders for the advantage of low calcination temperature and high homogeneity. The example illustrated in Figure 2.32 shows the procedures of gel-casting method for the synthesis of lanthanum strontium manganite powders, conducting by us[33].

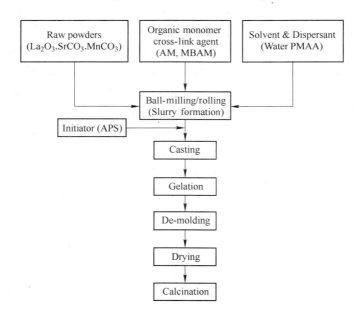

Figure 2.32 Procedures of the gel-casting method for the synthesis of LSM powders

The lanthanum strontium manganite powders are traditional cathode materials for solid oxide fuel cells. Both thermal and X-ray diffraction analysis confirmed that the gel-casting LSM powder formed a single perovskite phase at 850 ℃, which is 100 ~ 150 ℃ lower than that of the LSM powder prepared by the conventional solid-state reaction route. The significantly reduced phase formation temperature of the gel-casting LSM powder is most likely due to the homogeneously distributed and immobilized precursor particles in a polymeric network, promoting the sintering and crystallization process.

2.2.3 Powders by Vapor-phase Reactions

Vapor phase processes in general are of such growing importance in ceramic forming and processing, which is most often used to produce strongly adherent coatings on substrates. However, if conditions are chosen so that adhesion does not occur, then a fine powder with desirable morphology can be produced. Vapor phase processes are relatively expensive, but there are several good reasons for using them to prepare powders, particularly when we want: (1) High purity; (2) Discrete and nonaggregated particles; (3) Nanoparticles with narrow size distributions; (4) Versatility in producing powders of oxides and nonoxides[34].

Most synthesis methods of nanoparticles in the gas phase are based on homogeneous nucleation in the gas phase and subsequent condensation and coagulation. The ablation of a solid source with a pulsed laser can also yield nanoparticles, but the formation mechanism is at present not very clear. A micronsized aerosol droplet may also yield nanoparticles by evaporating a solute-containing droplet. A preview of the gas-phase synthesis methods is shown in Figure 2.33. In the following sections some important methods will be described.

2.2.3.1 Flame synthesis

Nanoparticles are produced by employing the flame heat to initiate chemical reactions producing condensable monomers. The flame route has the advantage of being an inexpensive method, however usually it yields agglomerated particles.

Flame synthesis of TiO_2 and SiO_2 form two of the largest industrial processes for synthesizing powders by gas phase reactions. The reactions can be written:

$$TiCl_4(g) + 2H_2O(g) \longrightarrow TiO_2(s) + 4HCl(g)$$
$$SiCl_4(g) + O_2(g) \longrightarrow SiO_2(s) + 2Cl_2(g)$$

and particle formation is illustrated in Figure 2.34. In the formation of fumed SiO_2, $SiCl_4$ reacts in an H_2 flame (about 1800 ℃) to form single spherical droplets of SiO_2.

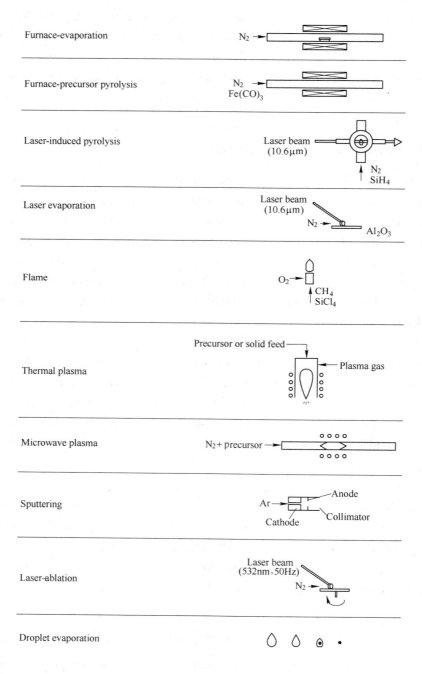

Figure 2.33 Schematic view of different synthesis methods[34]

These grow by collision and coalescence for to form larger droplets. As the droplets begin to solidify, they stick together on collision but do not coalesce, forming solid aggregates, which in turn continue to collide to form agglomerates.

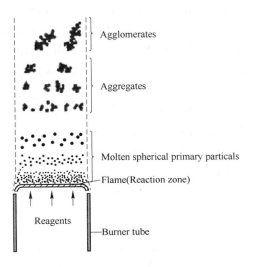

Figure 2.34   Schematic diagram illustrating the formation of primary particles, aggregates and agglomerates in gas phase reactions heated by a flame[8]

2.2.3.2 Plasma synthesis

A variety of ceramic powders of high purity and very small panicle size (10 to 20 nm) have been synthesized in high-temperature plasma environments. The particles essentially condense in a flowing gas, which accounts for the high purity. The plasma temperatures are in the order of 10^4 ℃, decomposing the reactants into ions and dissociating atoms and radicals. Solid powder feeds can also be decomposed by the plasma. Nanoparticles are formed upon cooling while exiting the plasma region. Two types of plasma reactors have been successfully used. One is the DC arc jet system. In this system the plasma is in direct contact with the metal electrode that supplies current. This reactor has very high efficiency, but can result in trace impurities from the electrode. The second type of plasma reactor is an rf (radio frequency) induction system. In this system, the current is transferred to the plasma through the electromagnetic field of the induction coil. No direct contact occurs, so no contamination results in the powder being synthesized. The efficiency of the rf induction system is lower than the DC arc jet system. However, both have produced SiC particles with greater than 70% efficiency using $SiCl_4$, CH_4, and H_2 as the gaseous precursors. Si_3N_4 has also been synthesized by plasma techniques.

2.2.3.3 Laser synthesis

In the laser pyrolysis technique, being a special class of laser processing techniques, a flowing reactant gas is heated rapidly with an IR laser such as a cw CO_2 laser. The source molecules are heated selectively by absorption of the laser beam energy, whereas the carrier gas is only indirectly heated by collisions with the reactant molecules. A gas phase decomposition of the reactants takes place due to the temperature increase and supersaturation is created.

A laboratory scale reaction cell is shown in Figure 2.35. The laser beam enters the cell through a KCl window and intersects the stream of reactant gases, usually diluted with an inert gas such as argon. The powders are captured on a filter located between the cell and a vacuum pump.

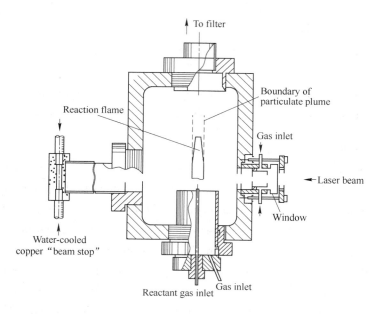

Figure 2.35 Laboratory scale reaction cell for the preparation of powders by laser heating of gases[8]

An advantage of the method is that the reactions can be fairly well controlled by manipulation of the process variables such as the cell pressure, the flow rate of the reactant and dilutant gases, the intensity of the laser beam, and the reaction flame temperature. The reactions by SiH_4 with NH_3, and SiH_4 with C_2H_4 have been used for the production of Si_3N_4 and SiC powders. An advantage of using SiH_4 rather than $SiCl_4$ as a reactant is that it has a strong adsorption band near the wavelength of the laser (10.6μm).

2.2.4 Spray Drying

The development of spray drying equipment and techniques evolved over a period of several decades from the 1870s through the early 1900s. Spray drying comes of age during World War II, with the sudden need to reduce the transport weight of foods and other materials. This technique enables the transformation of feed from a fluid state into dried particulate form by spraying the feed into a hot drying medium. It is a continuous particle processing drying operation. The feed can be a solution, suspension, dispersion or emulsion. The dried product can be in the form of powders, granules or agglomerates depending upon the physical and chemical properties of the feed, the dryer design and final powder properties desired[35].

In spray drying, a solution or a suspension is broken up into fine droplets by a fluid atomizer and sprayed into a drying chamber. Contact between the spray and drying medium (commonly hot air) leads to evaporation of moisture. The powder, consisting of dry particles, is carried out in the air stream leaving the chamber and collected using a bag collector or another form of collector (Figure 2.36).

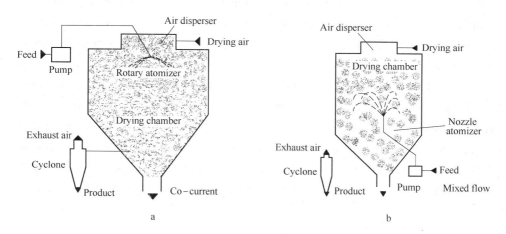

Figure 2.36 Spray dryers

a—Centrifugal atomizer with cocurrent air flow; b—Nozzle atomizer using mixed-flow conditions

The spray drying has two actions in advanced ceramic industry, first to dry ceramic slurry; second to dry and decompose the solution of metal salts (spray pyrolysis).

2.2.4.1 Spray pyrolysis

By using a higher temperature and a reactive (e.g., reducing) atmosphere in the chamber, solutions of metal salts can be dried and decomposed directly in a single step. This technique is referred to by many terms, including **spray pyrolysis, spray roasting, spray reaction**, and **evaporative decomposition of solutions**. Here we use the term spray pyrolysis. The idealized stages in the formation of a dense particle from a droplet of solution are shown schematically in Figure 2.37.

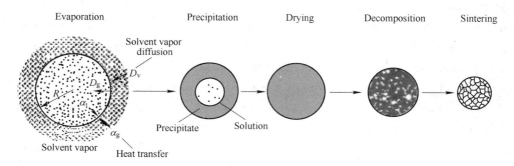

Figure 2.37 Schematic of the stages in the spray pyrolysis process[8]

The droplet undergoes evaporation and the solute concentration in the outer layer increases above the supersaturation limit, leading to the precipitation of fine particles. Precipitation is followed by a drying stage in which the vapor must now diffuse through the pores in the precipitated layer. Decomposition of the precipitated salts produces a porous particle made up of very fine grains, which is finally heated to produce a dense particle. In practice, a variety of particle morphologies can be produced in the spray pyrolysis process. For the fabrication of advanced ceramics, dense particles are preferred over those with highly porous or hollow shell-like morphologies because a subsequent milling step is normally unnecessary. An understanding of the key processing conditions is useful for achieving the desired particle characteristics. In addition to producing powders spray pyrolysis technique has been used to produce thin films and fibers.

2.2.4.2 Freeze drying

Freeze drying (also known as cryochemical processing) is a process that was first reported in 1968 by Schnettler et al[10]. It has potential for producing uniform particle and crystallite sizing of very pure, homogeneous powder. In freeze drying, a solution of metal salt is broken up by an atomizer into fine droplets, which are then frozen rapidly by being sprayed into a cold bath of immiscible liquid such as hexane and dry ice or directly into liquid nitrogen. The frozen droplets are then placed in a cooled vacuum chamber and the solvent is removed, under the action of a vacuum, by sublimation without any melting. The system may be heated slightly to aid the sublimation. The technique produces spherical agglomerates of fine primary particles with the agglomerate size being the same size as that of the frozen droplets. The size of the primary particles (in the range of 10 ~ 500nm) depends on the processing parameters such as the rate of freezing, the concentration of metal salt in the solution, and the chemical composition of the salt. After drying, the salt is decomposed at elevated temperatures to produce an oxide.

As being observed for spray drying, the breaking up of the solution into droplets serves to limit the scale of agglomeration or segregation to the size of the droplet. The solubility of most salts decreases with temperature and the rapid cooling of the droplets in freeze drying produces a state of supersaturation of the droplet solution very rapidly. Particle nucleation is therefore rapid and growth slow so that the size of the particles in the frozen droplet can be very fine. When compared to the evaporation of the liquid in spray drying, the approach to supersaturation is relatively faster so that freeze drying produces much finer primary particles with a higher surface area per unit mass. Surface areas as high as $60 m^2/g$ have been reported for freeze-dried powders.

Freeze drying of solutions has been used to prepare many kinds ceramic powders. Pressing of such powders produced fairly homogeneous green bodies.

2.2.4.3 Spray drying of ceramic slurry

Spray drying is commonly used in ceramic processing to achieve a uniform, free-flowing powder. As shown in Figure 2.38, in spray drying there are two kinds of atomizing types: centrifugal and nozzle (pressure). Figure 2.39 shows a pressure nozzle spray dryer (a) using in the production of alumina ceramics (Jingang New Materials Co. China) and a centrifugal spray dryer (b) using in the production of silicon carbide ceramics (Shandong Baona New Materials Co.).

In the alumina ceramic slurry drying, the slurry as mentioned in section 2.2.1.2, with water content: 34% ~ 35%, viscosity: >36s, D_{50}: 2.3 ~ 2.6μm, D_{90}: 4.5 ~ 5.2μm, is pumped to the middle part of the drying tower by a two-cylinder pump as shown in Figure 2.38a, atomized through the nozzles, then a thermal exchange between the droplets and hot air occurs, the alumina ceramic powders are collected in a bag under the drying tower.

The as-produced powders have a water content of 0.6% ~ 0.8% (mass fraction), packing density of 1.11 ~ 1.14g/cm^3, and the grain-size distribution of: >40 mesh:

Figure 2.38 Photographs of spray dryers
a—Pressure nozzle spray dryer; b—Centrifugal spray dryer

Figure 2.39 Morphologies of the sprayed powders
a—alumina ceramic powder; b—SiC ceramic powder

≤3.0%; 40~60 mesh: 20%~30%; 60~100 mesh: 50%~60%; 100~120 mesh: 5%~15%; 120~140 mesh: ≤10%; and <140 mesh: ≤5%. Figure 2.39a shows the morphology of the sprayed alumina ceramic powders.

During the production of reaction sintering SiC ceramics (RBSC) (Shandong Baona New Material Co.), the slurry is dried in a centrifugal spray dryer. The powders as-produced have to satisfy requirements of the processing of mold press: water content of 0.8wt% ~ 1.2wt%, packing density of 1.02 ~ 1.06g/cm^3, and the grain-size distribution of: >40 mesh: ≤5.0%; 40 ~ 60 mesh: 20% ~ 30%; 60 ~ 120 mesh 55% ~ 65%; and <120 mesh ≤5%. Figure 2.39b shows the morphology of the sprayed RBSC powders.

In the production of advanced ceramics, most spray drying is conducted with water as the carrier for the ceramic particles. However, some powders (especially nonoxides) react with water. Closed-cycle spray dryers utilizing alcohol or other non-aqueous fluid as media have to use. Now a closed-cycle spray drying has been used successfully with Si_3N_4 ceramic powders in the production of Si_3N_4 ceramic bearings (Institute of Zhongcai Artificial Crystals, China).

2.3 Characterizing Powders

There are several techniques that can be used to obtain particle size and particle size distribution. The choice of technique depends on several factors, such as applicable particle size range, sample size required, and the analysis time. In addition, the instrument cost, availability, ease of operation, and maintenance must be considered[25].

2.3.1 Characterizing Powders by Microscropy

The most direct way to determine the size of a particle is to look at it. If the size of the particle is >1μm, then visible light microscopy (VLM) is fine. Particle size measurements are made either directly at the microscope or from micrographs (photographs taken using the microscope). The main challenge is in determining the size of three-dimensional grains on the basis of planar images. Several procedures have been employed for making these measurements. The Heyn intercept method is one of the most useful approaches, and is ideally suited for nonequiaxed grains. The number of grain or grain boundary intersections of a straight or curved line is measured and from this information the grain size is determined. It is possible to make these measurements by hand using a ruler, but it would take a long time to obtain a statistically relevant sample. Using image analysis methods on a computer a large number of particles can be measured quickly. The data are often then plotted as a histogram of frequency of occurrence versus particle size.

For submicron particles it is necessary to use an electron microscope. For scanning electron microscopy (SEM), and in particular transmission electron microscopy (TEM), the total amount of material that can be examined is quite small, and so it is essential to make sure that the sample examined is representative of the entire powder batch.

The digital readout on a TEM is not more than ±10% accurate. To obtain more accurate measurements you must first calibrate the magnification of the instrument.

2.3.2 Characterizing Powders by Light Scattering

When a beam of light strikes a particle, some of it is transmitted, some is absorbed, and some is scattered. When the particles are larger than the wavelength of the incident light they cause Fraunhofer diffraction. The intensity of the forward-scattered light (i.e., light traveling in roughly the same direction as the incident light) is proportional to d^2. Figure 2.40 shows examples of the light scattered from two particles of different sizes. Smaller particles scatter a small amount of light through a large angle. Large particles scatter a greater amount of light but through a smaller angle. The relationship between scattering angle (θ) and d is: $\sin\theta = 1.22\lambda/d$.

The light source is usually an He-Ne laser with $\lambda = 0.63\,\mu m$. For this wavelength the reliable particle size range is $2 \sim 100\,\mu m$. Light-scattering methods have the advantages of accuracy, high speed, speediness, small sample size and being automated.

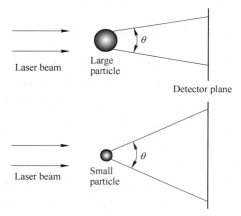

Figure 2.40 Scattering of light by large and small particles

2.3.3 Characterizing Powders by X-Ray Diffraction

X-ray diffraction (XRD) can be used to obtain the crystallite size. Because of the widespread use of this technique and its applicability to very small particles we will reiterate some of the key points as they apply to characterizing powders. The width of the diffraction peaks, β, is related to d by the Scherrer equation:

$$d = 0.9\lambda/(\beta\cos\theta)$$

where λ is the X-ray wavelength and θ is the Bragg angle. From above equation, we can see that as d increases, β decreases. When d is greater than about 0.1 mm the peaks are so narrow that their width cannot be distinguished from instrumental broadening. Consequently, XRD is most applicable to fairly small particle sizes. Figure 2.41 shows a series of XRD profiles for the 111 peak (arising from diffraction of the X-rays by the {111} planes) of a ZrO_2 powder doped with 3 mol% Y_2O_3. Higher calcination temperatures lead to particle coarsening and a corresponding decrease in β.

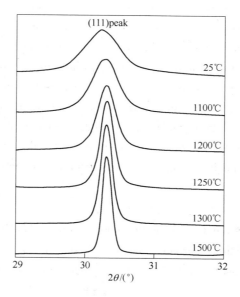

Figure 2.41 Illustration of X-ray fine broadening for a ZrO_2/3 mol% Y_2O_3 powder prepared by hydrothermal synthesis

It is important to remember that when determining particle size in a powder by measuring the width of X-ray peaks it is actually the size of the individual crystals that are being measured. As a consequence, if the particles are agglomerated, XRD will give the size of the primary particles and not the agglomerate size.

Similarly, the reflections (spots) in an electron diffraction pattern will be broadened if the sample is composed of small crystals. Therefore, diffraction in the TEM is not normally used to determine particle size because the number of particles that can be examined is fairly small and because it is better to just look at the image and make the measurements directly.

2.3.4 Characterizing Powders by Surface Area (The BET Method)

Surface-area methods rely on the adsorption of gases onto a particle surface at low temperature. The mass of gas adsorbed is measured as a function of gas pressure at a fixed temperature (typically liquid nitrogen). The method developed by Brunauer, Emmett, and Teller (BET) to estimate the particle size relies on determining the surface area of the powder, which is calculated from the N_2-isotherm observed at the boiling point of N_2.

The BET equation is:
$$p/[V_a(p_0 - p)] = (V_m C)^{-1} + (C - 1)p/[p_0 V_m C]$$

where p is the gas pressure; p_0 is the saturation vapor pressure for the adsorbate at the adsorption temperature; V_a is the adsorbate volume at relative pressure p/p_0; V_m is the adsorbate volume per unit mass of solid for monolayer coverage; C is the BET constant.

V_m is determined in the relative pressure p/p_0 which ranges from about 0.05 to 0.2; according to BET theory this is the amount of nitrogen necessary to form a monomolecular layer on the particle. Since one nitrogen molecule requires a surface area of 0.162 nm^2, the surface area of the particle can be easily estimated in m^2/g.

A plot of $p/[V_a(p_0 - p)]$ versus p/p_0 gives a straight line from which V_m and C can be determined. The specific surface area, S, of the powder can then be calculated using:
$$S = N_A \sigma V_m / V'$$

where N_A is Avogadro's number, V' is the molar volume = $22140 cm^3/mol$, and σ is the cross-sectional area of the adsorbate molecule ($0.162 nm^2$ for N_2).

For spherical particles the particle radius, a, can be obtained from:
$$a = 3/(\rho S)$$

where ρ is the density.

2.3.5 Determining Particle Composition and Purity

In addition to knowing particle size and particle size distribution of the powders, we often need to know its composition and purity. Industrial ceramic powders can contain over 30 detectable elements, but in most cases less than 10 are present at levels greater than 0.01% ~0.05%. Many industries use wet chemical techniques such as precipitation and titration for such analysis. These techniques are used because they are often simple to perform and give a quick result. For example, in the industrial production of red lead (Pb_3O_4) it is necessary to determine the amount of free Pb and litharge

(PbO). This analysis is typically done hourly and the results are used to modify the furnace temperature or throughput.

In addition to using wet chemistry there are numerous analytical methods that can give us chemical composition and impurity levels and these are summarized in Table 2.6.

The choice of technique depends on several factors:

(1) Type of material (is it readily soluble in common solvents, is the powder agglomerated);

(2) Amount of material (do we have milligrams or kilograms);

(3) Possible impurities (alkali metals, H, rare earths);

(4) Amount of impurities (ppm or percent);

(5) Availability and cost of instrument (do we need to use a national facility).

Table 2.6 Chemical analysis of powders[25]

Bulk techniques	Comments
Emission spectroscopy (ES)	Elemental analysis to the ppm level, frequently used for qualitative survey analyses, 5 mg powder sample
Flame emission spectroscopy (FES)	Quantitative analysis of alkali and Ba to the ppm level, ppb detectability for some elements, solution sample
Atomic absorption spectroscopy (AAS)	Industry standard for quantitative elemental impurity analyses; detectability to ppm level, solution sample
X-ray fluorescence (XRF)	Elemental analyses, detectability to $10 \times 10^{-4}\%$, $Z > 11$, solid/liquid samples
Gas chromatography/mass spectrometry (GC/MS)	Identification of compounds and analysis of vapors and gases
Infrared (IR) spectroscopy	Identification and structure of organic and inorganic compounds. Mg dispersed powder in transparent liquid of solid or thin-film sample
X-ray diffraction (XRD)	Identification and structure of crystalline phases, quantitative analysis to 1%, mg powder sample
Nuclear magnetic resonance (NMR)	Identification and structure of organic and inorganic compounds, sample to 5mg for H and 50mg for C

Of these factors, cost is often the most important. There are numerous choices:

(1) X-ray fluorescence (XRF) would not be a good choice to determine the amount of low-Z elements present.

(2) Flame emission spectroscopy (FES) is a good choice if we have very small

amounts of the alkali metals.

(3) Nuclear magnetic resonance (NMR) can be used to determine H concentrations, but it is often expensive to use and not as widespread as atomic absorption spectroscopy (AAS).

(4) For phase determination and phase proportions in a powder mixture XRD is useful, allowing quantitative phase analysis down to about 1% in a powder sample.

(5) With a field-emission source in TEM, chemical analysis with atomic resolution is possible; the interaction volume can be as small as about 10^{-8} mm^3.

Study Guide

2 – 1   What ceramic material is synthesized by the Acheson process?

2 – 2   Why is control of impurities in raw materials important for ceramics destined for high temperature, electrical magnetic or optical applications?

2 – 3   Why is control of particle size and particle size distribution important?

2 – 4   Why is powder "reactivity" important?

2 – 5   Explain the purpose or advantage of adding a milling aid during milling.

2 – 6   Identify some different processes that can yield highly controlled purity of powder to very fine particle size directly by chemical processes.

2 – 7   What are the reasons for spray drying or granulation?

References

[1] Aldinger F, Claussen N, Kaneno M, et al. Handbook of Advanced Ceramics: Materials, Applications, Processing, and Properties [M]. Academic Press, 2013, 90 ~ 106.

[2] Duval D J, Risbud S H, Shackelford J F. Mullite [M] //Ceramic and Glass Materials. Springer US, 2008: 27 ~ 39.

[3] Klug F J, Prochazka S, Doremus R H. Alumina-Silica Phase Diagram in the Mollite Region [J]. Journal of the American Ceramic Society, 1987, 70 (10): 750 ~ 759.

[4] Lee J H, Yonathan P, Yoon D H, et al. Dispersion Stability and its Effect on Tape Casting of Solvent-based SiC slurries [J]. J Ceram Process Res, 2009, 10: 301 ~ 307.

[5] Tong J F, Cheng D M, Li B W, et al. Microstructure and Performance of the Hot-Pressed Silicon Nitride Ceramics with Lu_2O_3 Additives [J]. Key Engineering Materials, 2010, 434: 37 ~ 41.

[6] Somiya, Shigeyuki. Advanced Technical Ceramics [M]. Elsevier, 1989.

[7] Liu N, Shi M, Xu Y D, et al. Effect of Starting Powders Size on the Al_2O_3-TiC Composites [J]. International Journal of Refractory Metals and Hard Materials, 2004, 22 (6): 265 ~ 269.

[8] Rahaman M N. Ceramic processing and Sintering [M]. Marcel Dekken Inc., 2003.

[9] Gock E, Kurrer K E. Eccentric vibratory mills—theory and practice [J]. Powder Technology, 1999, 105 (1): 302~310. SCHILLING R E, Yang M. Attritor grinding mills and new developments [J]. WESTERN COATINGS FEDERATION, 2000.

[10] David W Richerson. Modern Ceramic Engineering [M]. Third Edition. Taylor & Francis Group, 2006.

[11] Robert E S, Yang M, Attritor Grinding Mills and New Developments, Presented at Panamerican Coatings 2000, World Trade Center, Mexico City, Mexico, July 19, 2000.

[12] Ishll T, Hashimoto K. Effect of Dispersion Condition on Particle Size of Titanium Dioxide Dispersed by Bead Mill [J]. Journal of the Japan Society of Colour Material, 2012, 85 (4): 144~150.

[13] Baláž P. Mechanochemistry and Nanoscience [M] //Mechanochemistry in Nanoscience and Minerals Engineering. Springer Berlin Heidelberg, 2008: 1~102.

[14] Arthur L Robinson. A Chemical Route to Advanced Ceramics [J]. Science, 1986, 233 (4759): 25~27.

[15] Biswas D R. Development of novel ceramic processing [J]. Journal of materials science, 1989, 24 (10): 3791~3798.

[16] Rhine W E, Bowen H K. An overview of chemical and physical routes to advanced ceramic powders [J]. Ceramics international, 1991, 17 (3): 143~152.

[17] Edval G Araújo, Ricardo M Leal Neto, Marina F Pillis, et al. High-energy Ball Mill Processing. Third International Latin-American Conference on Powder Technology, November 26~28, 2001, Florianopolis, Brazil.

[18] Ristić M M, Milošević S D, Miljanić P. Mechanical Activation of Inorganic Materials [M]. SANU, 1998.

[19] Stojanovic B D. Mechanochemical Synthesis of Ceramic Powders with Perovskite Structure [J]. Journal of Materials Processing Technology, 2003, 143: 78~81.

[20] Lee S E, Xue J M, Wan D M, et al. Effects of Mechanical Activation on the Sintering and Dielectric Properties of Oxide-derived PZT [J]. Acta Materialia, 1999, 47 (9): 2633~2639.

[21] Xiaoyu Zhang, Yujun Zhang, Hongyu Gong, et al. Synthesis, Characterization and Optical Properties of $Y_{6-x}Sm_xMoO_{12+\sigma}$ Composite/Compounds Pigments with High Near-infrared Reflectance [J]. Advanced Materials Research, 2013, 602~604: 102~106.

[22] Riedel R, Dressler W. Chemical Formation of Ceramics [J]. Ceramics International, 1996, 22 (3): 233~239.

[23] Sorarù G D, Zhang Y, Ferrari M, et al. Novel Er-doped SiC/SiO_2 nanocomposites: Synthesis via polymer pyrolysis and their optical characterization [J]. Journal of the European Ceramic Society, 2005, 25 (2): 277~281.

[24] Livage J. Sol-gel processes [J]. Current Opinion in Solid State and Materials Science, 1997, 2 (2): 132~138.

[25] Carter Barry C and Norton Grant M. Ceramic Materials, Science and Engineering [M].

Springer Science + Business Media, LLC, 2007.

[26] Zhang Y, Quaranta A, Domenico Soraru G. Synthesis and luminescent Properties of Novel Eu^{2+}-doped Silicon Oxycarbide Glasses [J]. Optical Materials, 2004, 24 (4): 601~605.

[27] Liu X J, Zhang Y J. Synthesis and Neutron Absorption Properties of $Dy_2Ti_2O_7$ Nano-powders by Sol-gel Method, to be published.

[28] Wei R B, Zhang Y J, Gong H Y, et al. Pressureless Sintering of B_4C Neutron Absorbing Material with Nano-Sized Rare-Earth Compounds Addition [J]. Advanced Materials Research, 2013, 602: 503~507.

[29] Wei R B, Zhang Y J, Li X N, et al. Synthesis of a Dysprosium Aluminum Garnet Nanopowder via Sol-gel Method [J]. Journal of Sol-gel Science and Technology, 2013, 65 (3): 388~391.

[30] Byrappa K, Masahiro Yoshimura. Handbook of Hydrothermal Technology, A Technology for Crystal Growth and Materials Processing [M]. Noyes Publications, 2001.

[31] Namratha K, Byrappa K. Hydrothermal Processing and in Situ Surface Modification of Metal Oxide Nanomaterials. The Journal of Supercritical Fluids (2013), http: //dx. doi. org/10. 1016/j. supflu. 2013. 01. 007.

[32] Byrappa K, Adschiri T. Hydrothermal Technology for Nanotechnology [J]. Progress in Crystal Growth and Characterization of Materials, 2007, 53 (2): 117~166.

[33] Cheng C S, Zhang L, Zhang Y J, et al. Synthesis of $LaCoO_3$ Nano-powders by Aqueous Gelcasting for Intermediate Temperature Solid Oxide Fuel Cells [J]. Solid State Ionics, 2008, 179 (7): 282~289.

[34] Hlavacek V, Puszynski J A. Chemical Engineering Aspects of Advanced Ceramic Materials [J]. Industrial & Engineering Chemistry Research, 1996, 35 (2): 349~377.

[35] Patel R P, Patel M P, Suthar A M. Spray Drying Technology: An Overview [J]. Indian Journal of Science and Technology, 2009, 2 (10): 44~47.

3 Shape-Forming

Forming processes take a mix, slip, or plastic material and form it into a coherent, consolidated body having a chosen geometry. There are many processes available to perform this function. The selection of a forming operation for a particular product is very dependent on the size and dimensional tolerances of the product, the requisite microstructural characteristics, the levels of reproducibility required, economic considerations, and of course the required shape[1].

It is usually desirable to have a high green density as this controls the amount of shrinkage during firing. The microstructure of the green body has a significant effect on the subsequent firing stage. If severe variations in packing density occur in the green body, the fired body will, in general, contain heterogeneities that will limit the engineering properties.

Many forming methods are used for ceramic products and these can be grouped into three basic categories, which are not necessarily independent.

(1) Pressing: dry pressing, isostatic pressing, etc.
(2) Casting: slip casting, gel casting, tape casting, etc.
(3) Plastic forming: extrusion, injection molding, etc.

A flowchart illustrating the process of preparation for these three types is given in Figure 3.1.

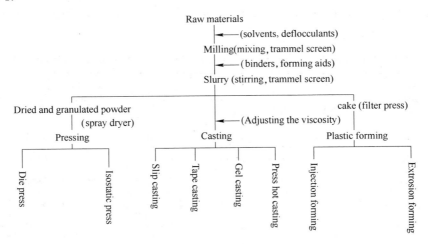

Figure 3.1 Flowchart illustrating the forming process

3.1 Additives

In the forming of ceramics, the use of certain additives is often vital for controlling the characteristics of the feed material, for achieving the desired shape, and for controlling the packing uniformity of the green body. The additives are either organic or inorganic in composition. Organic additives, which can be synthetic or natural in origin, find greater use in the forming of advanced ceramics because they can be removed almost completely prior to the sintering step. Therefore the presence of residues that can degrade the microstructure of the final product is largely eliminated. Inorganic additives cannot generally be removed after the forming step and are used in applications where the residues do not have an adverse effect on the properties of the final product[2].

The additives may be divided into four main categories: (1) solvents, (2) dispersants (also referred to as deflocculants), (3) binders, and (4) plasticizers. In some forming methods, other additives such as lubricants and wetting agents is also required.

3.1.1 Solvents

Solvents serve two major functions: (1) provide fluidity for the powder during forming; (2) dissolve the additives to be incorporated into the powder, thereby providing a means for uniformly dispersing the additives throughout the powder.

The solvents used in forming of advanced ceramics are water and organic liquids. Organic solvents generally have higher vapor pressure, lower latent heat of vaporization, lower boiling point, and lower surface tension than water, due in a large part to the strong hydrogen bonding of the water molecules. Table 3.1 listed the physical properties of some liquids at 20℃.

Table 3.1 Physical properties of some liquids (at 20℃ where applicable)[2]

Liquid	Density /g·cm^{-3}	Dielectric constant	Surface tension /N·m^{-1}	Viscosity /Pa·s	Latent heat of vaporization /kJ·g^{-1}	Boiling point/℃	Flash point /℃
Water	1.0	80	73×10^{-3}	1.0×10^{-3}	2.26	100.0	None
Methanol	0.789	33	23×10^{-3}	0.6×10^{-3}	1.10	64.6	18
Ethanol	0.789	24	23×10^{-3}	1.2×10^{-3}	0.86	78.4	20
Isopropanol	0.785	18	22×10^{-3}	2.4×10^{-3}	0.58	82.3	21
Acetone	0.781	21	25×10^{-3}	0.3×10^{-3}	0.55	56.0	-17
Methyl ethyl ketone	0.805	18	25×10^{-3}	0.4×10^{-3}	0.44	80	-1

Continues Table 3.1

Liquid	Density /g·cm^{-3}	Dielectric constant	Surface tension /N·m^{-1}	Viscosity /Pa·s	Latent heat of vaporization /kJ·g^{-1}	Boiling point/℃	Flash point /℃
Toluene	0.867	2.4	29×10^{-3}	0.6×10^{-3}	0.35	111	3
Xylene (ortho)	0.881	2	28×10^{-3}	0.7×10^{-3}	0.33	140	32
Trichloroethylene	1.456	3	25×10^{-3}	0.4×10^{-3}	0.24	87	None
n-Hexane	0.659	1.9	18×10^{-3}	0.3×10^{-3}	0.35	68.7	-23
Cyclohexanone	0.947	18	35×10^{-3}	0.8×10^{-3}	0.43	155	46
Mineral spirits	0.752					179~210	57

Water has a relatively high viscosity, and its tendency to form hydrogen bonds with hydroxyl groups on the surfaces of oxide powders can often steepen the effect of particle concentration on the suspension viscosity. The result is often a reduction in the solids content of the suspension for the maximum usable viscosity when compared to the use of an organic solvent such as toluene. However, as the understanding of surfactant chemistry and use of dispersants for aqueous media, the aqueous slurries can also reach a solid content of suspension which appears to be easier to achieve reproducibly in organic liquids.

For some powders (e.g., $BaTiO_3$, AlN, B_4C and Si_3N_4), their surfaces can be chemically attacked by water, leading to a change in composition and properties. For these powders, the use of an organic solvent is recommended. In order to prevent chemical attack of AlN in aqueous media, controlling the pH at a value of about 6 is available.

Although the disadvantages of water outlined above, problems with the disposal and toxicity of organic solvents are leading to a shift towards greater use of aqueous solvents. Two main areas of safety are flammability and toxicity. A commonly used indicator of flammability is the flash point which gives the temperature at which there is sufficient vapor, generated by evaporation, so that an already existing flame can cause a fire to start. Organic solvents such as toluene and methyl ethyle ketone, which are commonly used in tape casting, have very low flash points, so precautions must be taken to avoid explosions. Many organic liquids used in ceramic forming are toxic. Human exposure to these chemicals and waste disposal are important concerns. Trichloroethylene or a mixture of trichloroethylene and ethanol has been widely used in the tape casting industry because of its good solubility for many organic additives and inflammability. However, trichloroethylene and toluene are suspected of being carcinogens.

3.1.2 Dispersants

Dispersants, also referred to as deflocculants, serve to stabilize a slurry against flocculation by increasing the repulsion between the particles. The dispersant plays a key role in maximizing the particle concentration for some usable viscosity of the slurry.

Dispersants cover a wide range of chemical composition, and for many of them the composition is considered to be proprietary information by the manufacturers. The dispersants can be divided into three main classes, based on their chemical structure, as follows: (1) Simple ions and molecules; (2) Short chain polymers with a functional head (or end) group, commonly referred to as surfactants; (3) Low to medium molecular weight polymers.

3.1.2.1 Simple inorganic ions and molecules

Dispersants consisting of simple ions and molecules are effective in aqueous solvents. They are formed by dissociation of dissolved inorganic compounds such as salts, acids, and bases, often referred to as electrolytes. Examples are sodium silicate (Na_2SiO_3) commonly used for clays, tetrasodium pyrophosphate ($Na_4P_2O_7$), sodium hexametaphosphate ($Na_6P_6O_{18}$), sodium carbonate (Na_2CO_3), and hydrochloric acid (HCl) for oxides.

Preferential adsorption of one type of ions onto the particle surface coupledwith the formation of a diffuse layer of the counterions (ions of opposite charge) leads to electrostatic stabilization due to repulsion between the double layers. The valence and radius of the counterions can modify the repulsion between the particles and so can influence the stability of the suspension. Counterions with higher valence are more effective for causing flocculation (Schulze-Hardy rule), while for ions of the same valence, the smaller ions are more effective. For monovalent cations, the effectiveness of flocculation is in the order $Li^+ > Na^+ > K^+ > NH_4^+$, while for divalent cations, $Mg^{2+} > Ca^{2+} > Sr^{2+} > Ba^{2+}$. This sequence is known as the Hofmeister series. For common anions, the effectiveness of flocculation is in the order $SO_4^{2-} > Cl^- > NO_3^-$.

For advanced ceramics that must meet very specific property requirements, the use of inorganic dispersants may leave residual ions (e.g., sodium or phosphate), which even in very small concentrations can lead to the formation of liquid phases during sintering, thereby making microstructural control more difficult.

3.1.2.2 Short-chain polymers with a functional head group (Surfactants)

Surfactants contain a short-chain organic tail (containing up to 50~100 carbon atoms) and a functional head group that is nonionic or ionic in nature. For nonionic surfactants, the head group may be polar but does not ionize to produce charged species, so nonionic surfactants are commonly effective in organic solvents. Adsorption onto the particle surfaces occurs either by van der Waals attraction or, more effectively, by stronger coordinate bonding. Using the Lewis acid-base concept, an atom in the surfactant functional group which has an unshared pair of electrons (e.g., N or O) may act as a Lewis base and form a coordinate bond with an atom (e.g., Al) on the particle surface which has an incomplete shell of electrons (Lewis acid). Stabilization most likely occurs by steric repulsion between the organic tails or micelles that are stretched out in the organic solvent. In addition, menhaden fish oil is a widely used dispersant for Al_2O_3, $BaTiO_3$, and several other oxides. It consists of a mixture of several short-chain fatty acids with the alkyl chain containing some C=C double bonds and the functional end group being a carboxylic acid (COOH). Polyisobutylene succinamide (OLOA-1200) is commonly used for dispersing carbon particles in organic solvents.

Ionic surfactants are described as either anionic, when the functional head group ionizes to form a negatively charged species, or cationic, when a positively charged head group is formed (Figure 3.2a). They are effective in aqueous solvents. Usu-

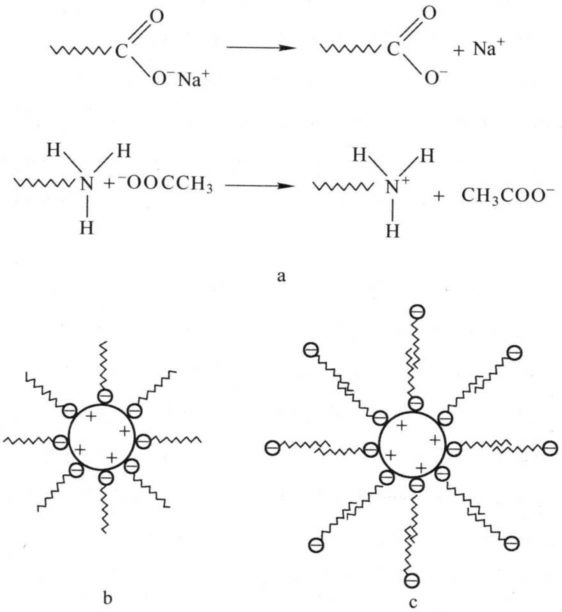

Figure 3.2 Dissociation of anionic and cationic surfactants to form negatively charged and positively charged head groups, respectively (a), Stabilization by negatively charged surfactant (b), and Stabilization by negatively charged micelles (c)

ally, negatively charged oxygen species are formed on dissociation of anionic surfactants. Adsorption of the surfactant commonly occurs by electrostatic attraction with positively charged particle surfaces. Stabilization of the suspension occurs essentially by electrostatic repulsion between the negative charges due to the adsorbed surfactant molecules (Figure 3.2b) or micelles (Figure 3.2c). Cationic surfactants commonly consist of positively charged nitrogen species on dissociation. Except for the reversal in the sign of the charges on the surfactant and the particle surface, the mechanisms of adsorption and stabilization are similar to the anionic surfactant case.

3.1.2.3 Low to medium molecular weight polymers

These dispersants, with a molecular weight in the range of several hundred to several thousand, are also classified into nonionic and ionic types. Common nonionic polymeric dispersants are poly (ethylene oxide) (PEO), or poly (ethylene glycol) (PEG), poly (vinyl pyrrolidone) (PVP), poly (vinyl alcohol) (PVA), polystyrene (PS), and block copolymers of PEO/PS. With a higher molecular weight, many of these polymers are effective as binders. When the chain segment contains OH groups or polar species the dispersants are effective in water; otherwise they are effective in organic solvents. In organic or aqueous solvents, adsorption of the polymers can occur by the weaker van der Waals bonding or more effectively by coordinate bonding. In aqueous solvents, hydrogen bonding can also produce very effective adsorption. Because they are uncharged, nonionic polymeric dispersants provide stabilization by steric repulsion.

Also referred to as **polyelectrolytes**, ionic polymeric dispersants consist of chain segments that carry ionized groups and are therefore effective in aqueous solvents. On dissociation, the ionized groups in the chain segment can produce negatively charged species (anionic polymers) or positively charged species (cationic polymers). Some common anionic polymers are given in Figure 3.3. The sodium or ammonium salts of the polyacrylic acids have been used successfully with aqueous slurries of several oxide powders and their use is increasing. For advanced ceramics, the use of the sodium salt of these acids is not recommended because residual Na ions, even in very small concentrations, can create problems for microstructural control during sintering. An example of a cationic polymer is poly (ethylene imine), which becomes positively charged in acidic conditions but remains an undissociated weak base in basic conditions.

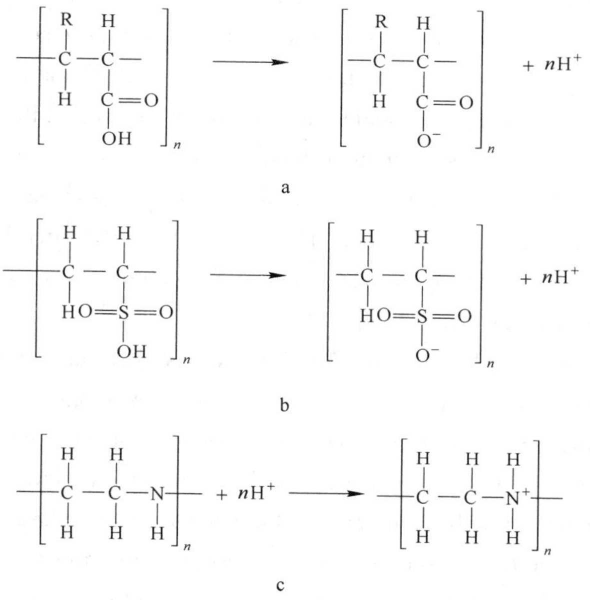

Figure 3.3 Examples of common anionic and cationic polymers used as dispersants
a—poly (acrylic acid): R=H; poly (methacrylic acid): R=CH$_3$;
b—poly (vinyl sulfonic acid); c—poly (ethylene imine)

3.1.3 Binder

Binder is a component that is added to hold the powder together while the body is formed and to provide strength to the green body. In some forming methods (e. g., injection molding), binders also provide plasticity to the feed material to aid the forming process. Table 3.2 listed a variety of organic and inorganic materials that have been used as binders.

Table 3.2 Examples of binders used in advanced ceramic processing[3]

Organic	Inorganic
Polyvinyl alcohol (PVA); Waxes; Celluloses; Dextrines; Thermoplastic resins; Thermosetting resins; Chlorinated hydrocarbons; Alginates; Lignins; Gums; Starches; Flours; Casein; Gelatins; Albumins; Proteins; Bitumens; Acrylics	Clays; Bentonites; Mg-Al silicates; Soluble silicates; Organic silicates; Colloidal silica; Colloidal alumina; Aluminates; Phosphates; Borophosphates

The clay minerals such as kaolinite are a good example of inorganic binders. Kaolinite has a layered structure and interacts with water to yield a flexible, plastic mixture. The clay minerals do not burn off during densification, but instead become part of the ceramic.

In the forming of advanced ceramic products, long chain polymers are widely used as organic binders. Some of organic binders are soluble in water, while others are soluble in organic liquids. The common synthetic binders are the vinyls, acrylics, and the ethylene oxides (glycols). The vinyls have a linear chain backbone in which the side group is attached to every other C atom. The acrylics have the same backbone structure but may have one or two side groups attached to the C atom. The cellulose derivatives are a class of naturally occurring binders[2].

In the selection of a binder for given forming process, several factors must be considered, which include binder burnout characteristics, molecular weight, compatibility with the dispersant, effect on the viscosity of the solvent, solubility in the solvent, and cost etc. It is clear that low cost is a key consideration in industry.

The organic binders as well as the other additives used to aid the forming of the green body must normally be burned out as completely as possible prior to sintering. The concentration of the binder is commonly much greater than that of the other additives, so the burnout characteristics is of primary importance in the selection of the binder. The burnout characteristics depend primarily on the binder chemistry and the firing atmosphere.

If a dispersant is used in the forming process, then the binder should be compatible with it. In general, the binder should not displace the dispersant from the particle surface which, for oxides, commonly means that the binder molecule should be less polar than the dispersant. The effect of the binder on the rheology of the solvent is a key problem. The organic binders increase the viscosity and change the flow characteristics of the liquid. Some can even lead to the formation of a gel. In the casting methods, the binder should not produce a rapid increase in the viscosity of the solvent with increasing concentration because this will limit the amount of powder that can be incorporated into the suspension for some usable viscosity. On the other hand, a rapid increase is generally desirable in extrusion to provide good green strength with a small concentration of binder[2].

The binder is commonly added as a solution in most of the forming methods, so its solubility in the liquid is an important factor. The backbone of the molecule consists of covalently bonded atoms such as carbon, oxygen, and nitrogen. Attached to the backbone are side groups located at frequent intervals along the length of the molecule. The chemical nature of the side groups determines, in part, what liquids will dissolve the binder. If binders have similar functional groups or similar molecular polarity, their solubility in the liquid would be enhanced[2].

3.1.4 Plasticizers

Plasticizers are generally organic substances with a lower molecular weight than the binder. The primary function of the plasticizer is to soften the binder in the dry state (i.e., reduce the T_g of the binder), thereby increasing the flexibility of the green body (e.g., tapes formed by tape casting). For forming processes in which the binder is introduced as a solution, the plasticizer must be soluble in the same liquid used to dissolve the binder. In the dry state, the binder and plasticizer are homogeneously mixed as a single substance. The plasticizer molecules get between the polymer chains of the binder, thereby disrupting the chain alignment and reducing the van der Waals bonding between adjacent chains. This leads to softening of the binder but also reduces the strength. Some commonly used plasticizers are listed in Table 3.3.

Table 3.3 Common plasticizers used in ceramic processing[2]

Plasticizer	Melting point/℃	Boiling point/℃	Molecular weight
Water	0	100	18
Ethylene glycol	-13	197	62
Diethylene glycol	-8	245	106
Triethylene glycol	-7	288	150
Tetraethylene glycol	-5	327	194
Poly (ethylene glycol)	-10	>330	300
Glycerol	18	290	92
Dibutyl phthalate	-35	340	278
Dimethyl phthalate	1	284	194

3.1.5 Other Additives

The number of additives used in a given forming process should be kept to a minimum because the potential for undesirable interactions between the components increases with their number. However, small amounts of other additives are sometimes used to serve special functions.

Wetting agents are surfactants added primarily to reduce the surface tension of liquids (particularly water), thereby improving the wetting of the particles by the liquid.

Lubricants are commonly used in die compaction, extrusion, and injection molding to reduce the friction between the particles themselves or between the particles and the die walls. Under the application of an external pressure, the particles rearrange more easily, leading to a higher and more uniform packing density. Common lubricants are steric acid, stearates, and various waxy substances.

Homogenizers such as cyclohexanone are sometimes used in tape casting to increase the mutual solubility of the components, thereby improving the homogeneity of the mixture.

3.2 Pressing

Pressing is accomplished by placing the powder into a die and applying pressure to achieve compaction. Two categories of pressing are commonly used: uniaxial and isostatic. Both use powder prepared by the same procedures[3]. For the thin object with simple shapes, an uniaxial pressure is usually applied, for more complex shapes isostatic pressing is required.

3.2.1 Uniaxial Pressing

Uniaxial pressing is ideally suited to the formation of simple solid shapes and consists of three basic steps: filling the die, compacting the contents, and ejecting the pressed solid. Most uniaxial presses are either mechanical or hydraulic. Mechanical presses typically have a higher production rate and are easy to automate.

Figure 3.4 shows schematically the pressing sequence of a typical uniaxial mechanical press. In a double-action press both the top and bottom punches are movable. The punches preposition in the die body to form a cavity predetermined (based on the compaction ratio of the powder) to contain the correct volume to achieve the required green dimensions after compaction. When the bottom punch is in the low position a cavity is formed in the die and this cavity is filled with free flowing powder. In dry pressing the powder mixture will contain between 0 and 5% (mass fraction) of a binder. (So, dry does not imply that there is no binder.) Once the cavity has been filled, the powder is struck off level with the top of the die. The top punch descends and compresses the powder either to a predetermined volume or to a set pressure. During pressing the powder particles must flow between the closing punches so that the space between them is uniformly filled. A particle size distribution of between 20 and 200 μm is often preferred for dry pressing: a high volume fraction of small particles causes problems with particle flow and also results in sticking of the punches. The pressures used in dry pressing may be as high as 300 MPa, depending upon material and press type, to maximize the density of the compact. After pressing, both punches move upward until the bottom punch

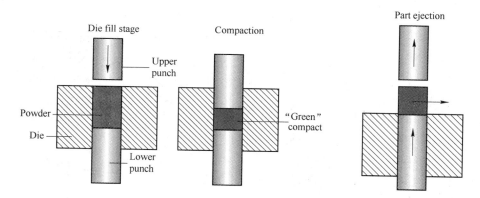

Figure 3.4 Schematic illustrating uniaxial pressing

is level with the top of the die and the top punch is clear of the powder-feeding mechanism. The compact is then ejected, the bottom punch is lowered, and the cycle is repeated.

Hydraulic presses transmit pressure via a fluid against a piston. They are usually operated to a set pressure, so that the size and characteristics of the pressed component are determined by the nature of the feed, the amount of die fill, and the pressure applied. Hydraulic presses can be very large, but have a much lower cycle rate than mechanical presses.

Following problems would be encountered in the uniaxial pressing: improper density or size; die wear; cracking; density variation.

The first two are easy to detect by simple measurements on the green compact immediately after pressing. Improper density or size are often associated with off-specification powder batch and are therefore relatively easy to resolve. Die wear shows up as progressive change in dimension, it should also be routinely handled by the process specification and quality control.

The source of cracking may be more difficult to locate. It may be due to improper die design, air entrapment, rebound during ejection from the die, die-wall friction, die wear, or other causes. Often a crack initiates at the top edge of the part during pressure release or ejection of the part.

Another important problem to be overcome in uniaxial pressing is nonuniform density. Density variation in the green compact causes warpage, distortion, or cracking during firing. One source of density variation is the friction between the powder and the die wall and between powder particles. As shown in Figure 3.5, a uniaxial pressure applied from

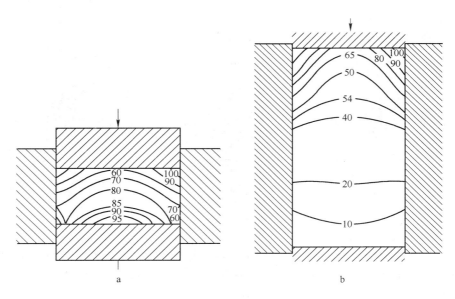

Figure 3.5 Pressure variations in uniaxial pressing due to die-wall friction and particle-particle friction, which lead to nonuniform density of the pressed compact[3]

a—$L/D = 0.45$; b—$L/D = 1.75$

one end of a die full of powder will be dissipated by friction so that a substantial portion of the powder will experience much lower than the applied pressure. These areas will compact to a lower density than the areas exposed to higher pressure. The pressure difference increases as the length-to-diameter ratio increases. During firing, the lower-density areas will either not density completely or will shrink more than surrounding areas.

Use of suitable binders and lubricants can reduce both die-wall and particle-particle friction and thus reduce density variation in the compact. Applying pressure from both ends of the die also helps.

A second source of nonuniform density is nonuniform fill of the die. Powder that is heaped otherwise nonuniformly stacked in the die will not reposition during pressing. The region with largest amount of powder will compact to a higher green density. This region of decreased porosity shrinks less during densification, resulting in distortion of the part.

A third source of nonuniform green density is the presence of hard agglomerates (cluster, particles) in the powder or a range of hardness of the granules in a free-flowing granulated powder. The hard granules will shield surrounding softer powder or granules from exposure to the maximum pressing pressure, resulting in pore clusters that reduce strength. Sometimes the surrounding powders will compact uniformly, but the hard agglomerate will trap porosity. The hard agglomerate may then shrink more than the surrounding material during densification and leave a large pore.

Note that the cause of nonuniform density is more associated with the condition of the powder loaded into the pressing die than with the pressing operation itself. Further, note that the problem may not show up until after the densification step of fabrication. This is another reminder that all of the processing steps are interlinked and that all must be coordinated and controlled to achieve the desired characteristics of the final product.

Figure 3.6 shows photos of presses. Figure 3.6a shows semiautomatic press made by Maichi Company, China. Figure 3.6b shows automatic press made by Laeis Company, Germany.

a   b

Figure 3.6 Photos of presses
a—Maichi 600/1200 hydraulic press; b—Laeis Alpha 1500 hydraulic press

3.2.2 Isostatic Pressing

The uniaxial pressing technique has its advantages, for example, the equipment for press forming is simple and the molding costs are low, then but the application of pressure in only one direction causes great friction between the powder and the mold walls, and transmission of the pressure within the power may suffer. Those factors can, depending on the shape of the object to be molded, lead to substantial unevenness in the transmission of pressure, which can cause distortions in shape and cracks when the object is fired. Therefore, the uniaxial mold press forming method is used only for molding the thin objects with comparatively simple shapes.

For molding the thick objects with more complex shapes, if applying pressure in all directions, some of the limitations that occur in uniaxial pressing can be overcome. Applying pressure from all directions is referred to as isostatic pressing or cold isostatic pressing (CIP). It has also been referred to as hydrostatic pressing.

Two types of isostatic pressing are commonly used: (1) wet-bag; (2) dry-bag.

3.2.2.1 Wet-Bag Isostatic Pressing

Wet-bag isostatic pressing in illustrated in Figure 3.7. The power is sealed in a watertight die. The walls of the die are flexible. The sealed die is immersed in a liquid contained in a high-pressure chamber. The chamber is sealed using a threaded or breach lock cover. The pressure of the liquid is increased by hydraulic pumping. The walls of the die deform and transmit the pressure uniformly to the powder, resulting in compaction. The walls of the die spring back after the pressure is removed, allowing the compact to easily be removed from the die after the cap of the die is removed[3].

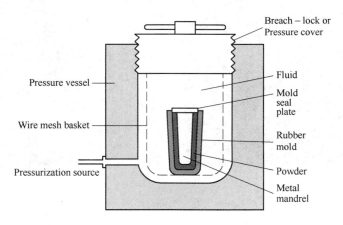

Figure 3.7 Schematic of a wet-bag isostatic pressing system[1]

Any noncompressible fluid can be used for isopressing. Water is commonly used, although fluids such as hydraulic oil and glycerine also work. The flexible walls of the die or mold are made of an elastomer such as rubber or polyurethane. The flexibility and wall thickness are carefully selected to allow optimum dimensional control and release characteristics. Natural rubber, neoprene, butyl rubber, nitrile, silicones, polysulphides, polyurethanes, and plasticized polyvinyl chloride have all been used.

Laboratory isostatic presses have been built with pressure capabilities among from 35 to 1380MPa. However, production units usually operate at 400MPa or less.

A major concern in isostatic pressing is uniform fill of the mold. This is usually achieved by use of vibration plus free-flowing spray-dried or granulated powder. Since higher pressures are usually achieved by isostatic pressing than by uniaxial pressing and since these pressures are applied uniformly, a greater degree of compaction is achieved. This usually results in improved densification characteristics during the subsequent sintering step of processing and a more uniform, defect-free component.

As with other processes, wet-bag isopressing has advantages and disadvantages. The advantages of the wet-bag process are:

(1) Wide range of shapes and sizes can be produced;
(2) Uniform density of the pressed product;
(3) Low tooling costs.

The disadvantages are:

(1) Poor shape and dimensional control (particularly for complex shapes);
(2) Products often require green machining after pressing;
(3) Long cycle times (typically between 5 and 60 minutes) give low production rates.

3.2.2.2 Dry-Bag Isostatic Pressing

Dry-bag isopressing was developed to achieve increased production rate and close dimensional tolerances. A schematic diagram of a mold for the dry-bag CIP is shown in Figure 3.8.

The main distinction of the dry-bag process is that the rubber mold is now an integral part of the press. The high-pressure fluid is applied through channels in the mold. After pressing, the pressed part is removed without disturbing the mold. Hence, the dry-bag press can be readily automated. Fully automated units are widely available and have been operating in the high volume production of ceramic parts for over 20 years. Production rates of up to 1 part per second are being achieved industrially. The dry-bag CIP has been used for many years to press spark plug insulators. The steps in this process are shown in Figure 3.9.

Figure 3.10 shows photos of isostatic presses. Figure 3.10a shows 500/1500/300 wet-bag isostatic press made by Xinkaiyuan, China. Figure 3.10b shows dry-bag isostatic press for producing of alumina balls in Jingang New Materials Co.

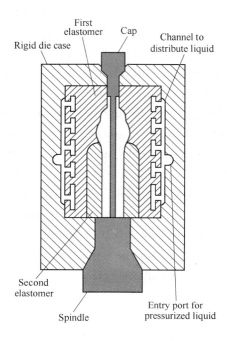

Figure 3.8 Schematic of a die for dry-bag isostatic pressing of a spark plug insulator[3]

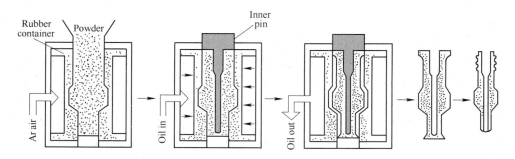

Figure 3.9 Making spark plugs[1]
(Hydrostatic pressure is applied by pumping oil around the rubber container,
which is part of the press and thus easily removed)

3.2.2.3 Green machining

As mentioned above, the products often require green machining after isopressing. Figure 3.11 shows photos of two lathes for green bodies machining, Figure 3.11a for Al_2O_3 ceramic tubes using in Jingang New Materilas Co. and Figure 3.11b for reaction sintering SiC sealing parts using in Baona New Materials Co.

a   b

Figure 3.10 Photos of isostatic presses
a—wet-bag isostatic press; b—Dry-bag isostatic press

a   b

Figure 3.11 Green machining lathes
a—a simply equipped for machining Al_2O_3 ceramic tubes; b—a numerically
controlled for machining RBSC sealing parts

For machining the green bodies pressed by isopressing process, the speed must be controlled. It is better that the green bodies of the reaction sintering SiC sealing parts are worked after drying for 2~3h, at about 100℃.

3.3 Casting

The casting of ceramics is similar to metal casting in which a shape is formed by pouring molten metal into a refractory mold. A limited amount of casting of molten ceramics is done in the preparation of high-density Al_2O_3 and Al_2O_3-ZrO_2 refractories and in preparation of some abrasive materials. The technique of casting molten ceramic refractories is called fusion casting.

More frequently, the casting of ceramics is done by a room-temperature operation in which ceramic particles suspended in a liquid are cast into a porous mold that removes the liquid and leaves a particulate compact in the mold. There are a number of variations to this process, depending on the viscosity of the ceramic-liquid suspension, the mold, and the procedures used. The common casting methods are slip casting, tape casting, gel casting and electrophoretic deposition etc.

The ideal requirements for the green body are homogeneous particle packing and a green density that is as high as possible. The slurry that is used to produce these characteristics must have a high particle concentration but must also have the rheological properties to flow easily during the casting process and to allow a high enough casting rate for economical production. The attainment of the slurry characteristics in a controlled manner requires an understanding of (1) the colloidal interactions between particles in a suspension and (2) the factors that control the rheological behavior of suspensions. An understanding of particle packing concepts and the use of organic additives in processing are also important.

The casting methods have the capability for producing a fairly homogeneous particle packing in the green body, but they are generally limited to the production of relatively thin articles. Slip casting offers a route for the production of complex shapes and is widely used in the traditional clay-based industry, for example, for the manufacture of pottery and sanitary ware. It has been steadily introduced over the past 50 years to the production of advanced ceramics[2].

3.3.1 Slip Casting

Pour slurry into a microporous plaster of Pairs ($2CaSO_4 \cdot H_2O$) mold; leave it for a time, and the water content in the slurry will be absorbed by the mold. Then the excess slurry is poured off, the molded object is dried in its mold, and then, when it is dry and strong, it is removed from the mold. **This process is known as slip casting**.

Slip casting as a forming process for clay wares was originated between the years 1700 and 1740. The historical development of the slip casting process as applied to sanitary ware and the theories to understand the casting behaviour have been reviewed by Rowlands[4]. Until 1910, the process of slip casting was limited to clay based materials. Alumina was the first non-clay material to be slip-cast according to Rado[5]. Thereafter the casting technique has been employed successfully for casting various oxides like silica, magnesia, zirconia, calcia, thoria etc. The successful casting of non-oxide materials including refractory metals, cemented carbides, nitrides and borides was explored in 1936. Since then, slip casting has emerged as one of the major forming techniques for large scale fabrication of both monolithic as well as composite advanced ceramic components of either very simple or very complicated shapes[6~10].

Figure 3.12 identifies the critical process steps in slip casting and some of the process

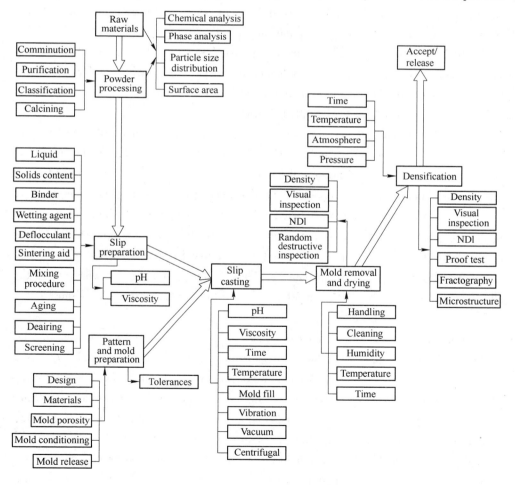

Figure 3.12 Critical process steps in slip casting[3]

parameters that must be carefully controlled to optimize strength or other critical properties.

The mold for slip casting must have controlled porosity so that it can remove the fluid from the slip by capillary action. The traditional mold material has been plaster. Some newer molds, especially for pressure casting, are made of a porous plastic material.

3.3.1.1 Plaster mold

Plaster of Paris is a mineral product made from calcined gypsum in which one half of a molecule of water is left in the structure. If lacking all the water, the material becomes anhydrous and is useless for casting.

The plaster consistency refers to parts of water per 100 parts of plaster by weight. For slip casting, a consistency of 60 ~ 70 is in the right range. As more water is added, the plaster strength decreases and absorption increases.

To obtain a consistency of 67, mix two parts of water to three parts of plaster by weight. This results in 0.065 final pounds of plaster per cubic inch. The formula establishes how much to mix provided the volume of the mold is known.

3.3.1.2 Drain casting and Solid casting

In ordinary slip casting, there are two methods: **drain casting** and **solid casting**. In drain casting, the cast wall is built up to the desired thickness and then the remaining slip is poured out. One can measure the wall thickness with a probe. This method is not too accurate but is often good enough for a lab procedure. If more accuracy is needed, one can use appropriate micrometers for this type of measurement. Solid casting is a procedure where the slip is cast into a solid part, just like the name implies. The basic approach of drain casting and solid casting is illustrated in Figure 3.13.

For the slip casting of reaction sintering SiC ceramic products, the viscosity of the slip needs to be controlled between 30 ~ 45s, and the absorption time 8 ~ 18min.

3.3.1.3 Pressure casting

One limitation with most slip-casting processes is the long time required to cast articles in the mold. This results in a large inventory of molds, high labor, and large floor space, all of which add to cost. Application of pressure to the slip increases the casting rate. This is referred to as **pressure casting**. It is similar to filter pressing. In filter pressing water or other liquid is removed from a powder by pressing the powder or liquid mixture against a permeable membrane. In pressure casting the slip is pressed into a shaped permeable mold. A schematic of the main features of a pressure casting device is shown in Figure 3.14. Original pressure casting was conducted with plaster molds. However, because of the low strength of the plaster, the amount of pressure that could

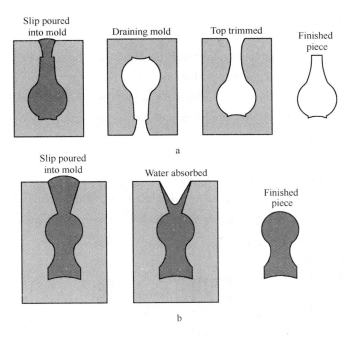

Figure 3.13 Schematic diagrams of drain casting (a) and solid casting (b)

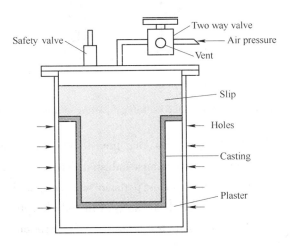

Figure 3.14 Pressure casting apparatus[2]

be applied was limited. Development of porous plastic molds allowed the pressure to be increased by ten times of the values of 3 ~ 4 MPa.

In order to improve the density of the body and save the time of casting process, besides the pressure casting, some other casting methods have been also used, for example, vacuum casting and centrifugal casting.

3.3.1.4 Vacuum casting

Vacuum casting can be conducted either with the drain or solid approach. A vacuum is pulled around the outside of the mold. The mold can consist of a rigid permeable form or of a thin permeable membrane (like filter paper) lining a porous rigid form. Vacuum casting is commonly used in the production of porous refractory fiberboard for lining high-temperature furnaces.

3.3.1.5 Centrifugal casting

Centrifugal casting involves spinning the mold to apply greater than normal gravitational loads to make sure that the slip completely fills the mold. This can be beneficial in the casting of some complex shapes. A schematic diagram of centrifugal casting of a ceramic membrane tube is shown in Figure 3.15.

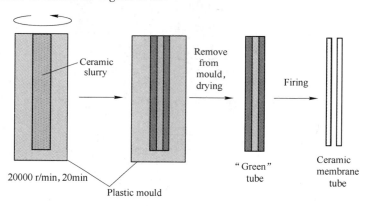

Figure 3.15 Schematic illustrating the centrifugal casting of a ceramic membrane tube

Centrifugal casting has been used for a long time in the metals and plastics industries. There is some literature in the ceramics industry. The advantages are that it can consolidate a part quickly, can make many parts in one cast, and can make intricate shapes with an appropriate mold. The disadvantages include keeping the apparatus balanced, cost of the equipment, and particle segregation. Segregation is a serious problem even with fine-grained slips[2].

3.3.1.6 Fugitive wax slip casting

The more complex shape can be also achieved using the soluble-mold casting technique. This is a relatively new approach based on the much older technology of investment casting. It is also referred to as **fugitive wax slip casting** and is accomplished in the following steps:

(1) A wax pattern of the desired con Figureuration is produced by injection molding a water soluble wax.

(2) The water-soluble wax pattern is dipped in a nonwater-soluble wax to form a thin layer over the pattern.

(3) The pattern wax is dissolved in water, leaving the nonwater-soluble wax as an accurate mold of the shape.

(4) The wax mold is trimmed, attached to a plaster block, and filled with the appropriate casting slip.

(5) After the casting is complete, the mold is removed by dissolving in a solvent.

(6) The cast shape is dried, green-machined as required, and densified at high temperature.

3.3.2 Tape Casting

With the type of device sketched in Figure 3.16, it is possible to coat a carrier tape or film with a slurry to a certain thickness, then strip off the carrier tape after drying and hardening, thereby producing the sheet-formed ceramic body. This process is known as tape casting, sometimes referred to as doctor-blade process. The overall process steps for producing the alumina ceramic tapes have been summarized in the flow chart in Figure 3.17.

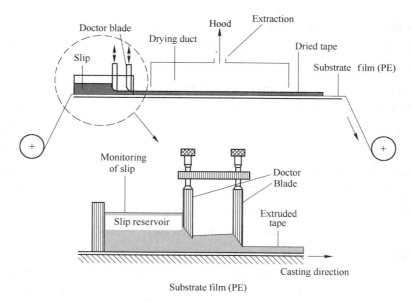

Figure 3.16 Schematic illustrating the tape-casting process[11]

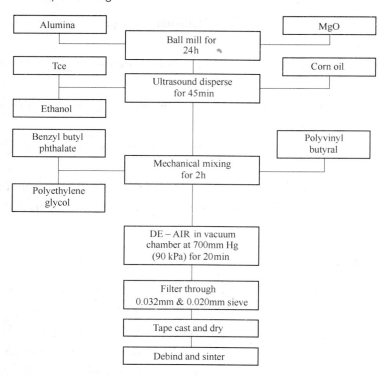

Figure 3.17　Tape casting flowchart[12]

  The technique consists of casting a slurry onto a moving carrier surface (usually a thin film of cellulose acetate, Teflon™, Mylar™, or cellophane) and spreading the slurry to a controlled thickness with the edge of a long, smooth blade. The slurry contains a binder system dissolved in a solvent. Enough binder is present so that a flexible tape will result when the solvent is removed. Solvent removal is achieved by evaporation. As with slip casting, the fluid must be removed slowly to avoid cracking, bubbles, or distortion. This is the purpose of the long portion of the tape-casting apparatus between the doctor blade and the take-up reel. The evaporation is achieved either by controlled heating or airflow. The dry flexible tape is rolled onto a reel to be stored for use.

  By selection of the appropriate binder and plasticizer for mixing into the slurry, it is possible to give the green tape characteristics that make it easy to process by punching, lamination, and so on. The preparation of the slurry is a critical step in the tape casting process. Most tape casting operations currently use organic solvents, but the trend is towards aqueous-based systems. Other considerations in the selection of a solvent are the thickness of the tape to be cast and the surface on which the cast is to be made. Thin tapes are cast from highly volatile solvent systems (e. g. , acetone or methyl ethyl ketone), whereas thicker tapes (about 0.25mm) have to be cast from slower drying solvents (e. g. , toluene).

The dispersant may be the most important organic additive in that it serves to lower the viscosity of the slurry, thereby allowing the use of a high particle concentration. Another important selection to be made is that of the binder-plasticizer combination because the concentration used in tape casting slurries is high. It must provide the required strength and flexibility of the green tape and must also be easily burnt out prior to sintering of the tape. Many organic systems can easily satisfy this criterion if the binder burnout process is carried out in an oxidizing atmosphere at reasonably high temperatures. However, several ceramic systems require the use of a binder-plasticizer system that can be removed in a nonoxidizing atmosphere. Typical slurry formulations for tape casting are given in Table 3.4.

The advantages and disadvantages of this process can be summarized as follows.

Advantages
Large-area, thin, flat parts
High densities achievable
Suitable for mass production
Economical
Green tape easily punched/cut
Mature technology

Disadvantages
Slurry composition dependent on many factors
Drying conditions must be carefully controlled
Environmental concerns

Table 3.4 Examples of tape-casting compositions[2] (mass fraction, %)

Powder	Solvent	Binder	Plasticizer	Dispersant	Other additives
Nonaqueous formulation for use in oxidizing atmospheres					
Al_2O_3 (59.5) MgO (0.1)	Ethanol (8.9) TCE (23.2)	PVB (2.4)	Octylphthalate (2.2) PEG (2.6)	Fish oil (1.0)	
Nonaqueous formulation for use in nonoxidizing atmospheres					
$BaTiO_3$ (69.9)	MEK (7.0) Ethanol (7.0)	30wt% solution of acrylic in MEK (9.3)	PEG (2.8) Butyl benzyl phthalate (2.8)	Fish oil (0.7)	Cyclohexanone (homogenizer) (0.5)
Aqueous formulation					
Al_2O_3 (69.0)	Deionized water (14.4)	Acrylic emulsion (cross-linkable) (6.9)	Acrylic emulsion (low T_g) (9.0)	Ammonium polyacrylate (0.6)	Poly (oxyalkylene-diamine) (0.1)

3.3.3 Gel casting

Gel casting was developed by combining traditional slip processing with polymer chemistry[13]. In gel casting, a concentrated slurry of ceramic powder in a solution of organic monomers is poured into a mold and then polymerized in situ to form a green body in the shape of the mold cavity. The monomer solution provides a low viscosity vehicle to transport the fluid slurry into the mold and the polymer gel holds the ceramic powder in the desired shape. This process separates the mold filling step from the setting of the mix. Vinyl monomers are used in the process and, because they undergo a free-radical chain polymerization reaction, the setting is very rapid. The vehicle removal is in two steps: drying to remove the solvent (water) followed by pyrolysis to remove the polymer. The part is then fired to densify it.

The initial gel casting process was developed by dissolving multifunctional acrylate monomers in organic solvents. These monomers when polymerized by free-radical initiators formed highly crosslinked polymer-solvent gels. When ceramic powder was slurried in this organic solution and cast into a mold and polymerized, the gel held the powder in the shape of the cavity of the mold. The green body was air-dried, the binder was burned off, and the part was sintered to full density. The gel casting concept was successfully demonstrated. In anticipation of environmental problems and the additional costs of the removal of the organic solvent, and because most ceramists prefer to work in water, an effort was begun to use water as the solvent.

The search for water-soluble monomers led to the acrylamide gel system which is used in biotechnology for gel electrophoresis. The acrylamide system produced excellent results, and became the standard gel casting system. Aqueous gel casting, where the solvent is water, and nonaqueous gel casting, where the solvent is organic had, thus, been successfully developed. The initial detailed studies of aqueous gel casting were carried out using alumina. The acrylamide system established aqueous gel casting as a major advanced ceramic forming process and demonstrated some of its advantages. Figure 3.18 shows the detailed flowchart of the gel casting process which is quite similar to a standard slip process. Instead of the usual polymer binder system, the binders here are monomers and initiator/catalyst is added later to effect the in situ polymerization.

It is important to distinguish gel casting from the sol-gel process. In gel casting, a high solids loading of a commercial powder in a solution of organic monomers is poured into a mold and then polymerized into an organic gel which holds the powder together. In sol-gel, the solid is usually produced as part of the process in the form of an inorgan-

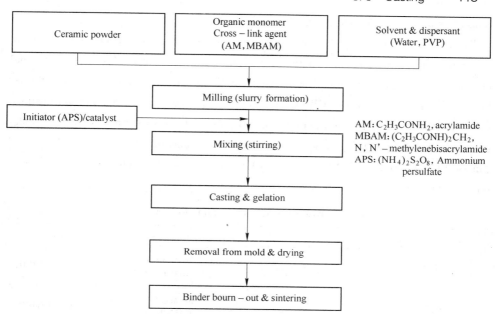

Figure 3.18 Detailed flowchart of the gelcasting process

ic gel with low solids loading.

It is worth to note the following key aspects of gel casting.

3.3.3.1 Premix solution

The premix solution is the solution of organic monomers in which the ceramic powder is suspended. It usually consists of a solvent which determines if the process is aqueous or nonaqueous gel casting, and two types of organic monomer. The main monomer has a single double bond and, if polymerized alone, forms a linear polymer. The second monomer is the crosslinking monomer which has at least two double bonds. When they are polymerized together, a crosslinked, polymersolvent gel results. Commonly used chain formers are methacrylamide (MAM), hydroxymethacrylamide (HMAM), N-vinyl pyrrolidone (NVP), and methoxy poly (ethylene glycol) monomethacrylate (MPEGMA). Sometimes two monomers may be used in combination (e.g., MAM and NVP, or MPEGMA and HMAM). The crosslinkers commonly used are methylene bisacrylamide (MBAM) and poly (ethylene glycol) dimethacrylate (PEGDMA). The free radical initiator used most often is ammonium persulfate (APS) with tetramethylethylene diamine (TEMED) used as a catalyst. In the acrylamide system the monomer is acrylamide (**AM**) and the crosslinker is N, N′-methylenebisacrylamide (**MBAM**); the standard premix solution contains 15% (mass fraction) total monomers in water with the mass ratio of **AM** to **MBAM** set at 24 ∶ 1.

The choice of the monomer system depends on several factors, such as the reactivity of the system (including the reaction temperature), the strength, stiffness and toughness of the gel, and the strength and machinability of the green body. One system used for several ceramics is MAM-MBAM in which the total monomer concentration in the solution is 10% ~20% (mass fraction) and the MAM to MBAM ratio is 2 ~6. Another system is MAM-PEGDMA in which the monomer concentration is also 10% ~ 20% (mass fraction) but the MAM to PEGDMA ratio is 1 ~3. Examples of gel casting compositions are given in Table 3.5.

Table 3.5　Examples of compositions for gel casting[2]

Ceramic powder	Dispersant	Monomer solution	Initiator
Al_2O_3	Ammonium polyacrylate	MAM-MBAM or MAM-PEGDMA	APS/TEMED
Si_3N_4	Poly (acrylic acid)	MAM-MBAM or MAM-PEGDMA	APS/TEMED
SiC	Tetrmethyl ammonium Hydroxide (pH11)	MAM-MBAM or MAM-PEGDMA	APS/TEMED

The formation of the gel occurs in two stages: initiation and polymerization. In the initiation stage, the viscosity does not change and no heat is generated. The addition of the initiator at room temperature allows a reasonable time (30 ~ 120min) for de-airing of the slurry and mold filling. The polymerization step is often conducted at an elevated temperature (40 ~80℃), and the faster reaction rate produces gelation in a short time. Because the reaction is exothermic, its progress can be followed by monitoring the temperature of the system.

3.3.3.2　Low-toxicity monomers

Although acrylamide gels have been used for decades in biology for gel electrophoresis in DNA and other analyses, and the polymer itself is safe, the acrylamide monomer is neurotoxic. A concerted research effort was undertaken to find less toxic monomers to replace acrylamide. With the assistance of an industrial hygienist, several monomers were examined based on their environmental, safety and health implications. The most important factor was the health hazard rating (4 = Extreme, 3 = High, 2 = Moderate, and 1 = Low). Acrylamide, which has the highest health hazard rating of 4, has been replaced by methacrylamide with a moderate rating of 2. The crosslinking monomer is either the **MBAM** (rating = 2) or poly (ethylene glycol) dimethacrylate (rating = 1).

3.3.3.3 Rheology of gel casting slurries

In gel casting, the solids loading of the slurry becomes the green density of the cast part. Consequently, it is important to have as high a solids loading as possible in the slurry. For advanced ceramics, to have uniform grain growth and acceptable final properties, narrowly dispersed submicron powders are used in the suspensions. In addition, high solids loading minimizes shrinkage and warpage during drying and enhances high sintered density. Therefore, in gel casting, it is desirable to have a suspension with at least 50% (volume fraction) solids which is fluid and pourable.

3.3.3.4 Drying of gel cast parts, binder burnout, and sintering

The drying of gel cast parts was investigated in order to reduce the drying time, the slowest step in the gel casting process. There was no constant rate drying period at any drying temperature. To avoid warpage and cracking caused by unrelieved stresses in the body, the drying of gel cast ceramics had to be at high relative humidities. Furthermore, it was found that shrinkage stopped early in the drying process when particle-to-particle contact had been established while the body continued to dry for a much longer time. Therefore, to decrease the drying time while minimizing unrelieved stresses, the initial drying is done at high relative humidities (>90%) until shrinkage has stopped. Then the drying rate is increased either by raising the drying temperature, or decreasing the relative humidity, or a combination of both.

The drying is isotropic and the linear shrinkage for typical 50% (volume fraction) solids loading is about 3%. Shrinkage was much less for higher solids loading and is negligible at about 70% (volume fraction) solids loading. After drying, the binder left in the green body is typically less than 4% (mass fraction).

The investigation of the binder burnout in air using thermogravimetric and differential thermal analyses (TGA/DTA) shows that all the binders come off primarily as complete combustion products, CO_2 and H_2O. The pyrolysis of the polymer is complete for Al_2O_3, green parts at below 600℃. Pyrolysis in N_2 leaves a residue of about 6% (mass fraction) of the polymer.

Gel cast parts were readily sintered to full density. The controlling parameter is the initial solids loading which should be high enough to produce an acceptable green density.

3.3.3.5 Mold materials

The commonly used mold materials for gel casting are aluminum, glass, polyvinylchloride, polystyrene, and polyethylene. Aluminum and especially anodized aluminum are used widely for permanent production molds, while glass and the polymeric materials are useful for laboratory experiments. The gel-casting system can react with the contact surfaces of the mold so the mold surfaces are often coated with mold release agents, such as the commercial mold releases employed in the polymer processing industry[2].

3.3.4 Electrophoretic deposition

Electrophoresis is the movement of charged particles through a liquid under the influence of an external electric field. The method of electrophoretic deposition (EPD) is shown schematically in Figure 3.19. A DC electric field causes the charged particles in a colloidal suspension to move toward and deposit on the oppositely charged electrode. EPD involves a combination of electrophoresis and particle deposition on the electrode. Successful EPD to form a deposit with high packing density requires a stable suspension. Agglomerated particles in an unstable suspension move toward the oppositely charged electrode and form a low-density deposit.

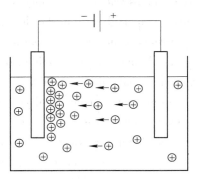

Figure 3.19 Schematic drawing of an electrophoretic deposition cell showing the process[14]

EPD is a facile forming technique that has the additional benefit of controllability because the rate and extent of deposition are manipulated electrically. It is best used for depositing coatings and thin objects.

Whereas electrophoresis is well understood, the deposition mechanism of the particles on the electrode has been the subject of much controversy, and several explanations have been put forward to explain the phenomenon. One explanation is illustrated in Figure 3.20. As a positively charged oxide particle with its surrounding double layer of counterions (the lyosphere) moves towards the cathode in an EPD cell, the electric field coupled with motion of the charged particle through the liquid causes a distortion of

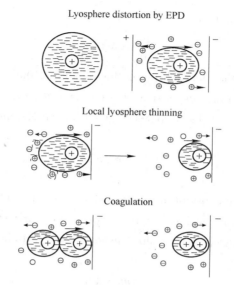

Figure 3.20 Schematic illustration of the electrophoretic deposition mechanism by lyosphere distortion and thinning[2]

the double-layer envelope: the envelope becomes thinner ahead of and thicker behind the particle. Compared to a particle far away from the electrode, the ξ-potential is greater for the leading hemisphere of the particle and smaller for the trailing hemisphere. Cations in the liquid also move to the cathode with the positively charged particles. The counterions in the trailing lyosphere tails tend to react with the high concentration of cations surrounding them, leading to a thinning of the double layer around the trailing surface of the particle. The next incoming particle, which has a thin double layer around the leading surface, can approach close enough so that van der Waals attractive forces dominate. The result is coagulation and deposition on the electrode.

The kinetics of EPD is important for controlling the thickness of the deposited layer. There are two modes of operating the system. In constant-voltage EPD, the voltage between the electrodes is maintained constant. Because deposition requires a steeper electric field than electrophoresis, as the deposition thickness (and consequently the electrical resistance) increases, the electric field decreases. The particle motion and, hence, the rate of deposition decreases. Under constant-current EPD, on the other hand, the electric field is maintained constant by increasing the total potential difference between the electrodes, so the limited deposition in constant-voltage EPD is avoided.

Assuming that the suspension is homogeneous and the change in concentration is due to EDP only, the mass of particles m deposited on the electrode is equal to that removed from the suspension; therefore:

$$\frac{dm}{dt} = AvC \tag{3.1}$$

where A is the area of the electrode; v is the velocity of the particle; C is the concentration of particles in the suspension; t is the deposition time. For a concentrated suspension, the velocity of the particles is given by the Helmholtz-Smoluchowski equation:

$$v = \frac{\varepsilon\varepsilon_0\zeta E}{\eta} \tag{3.2}$$

where ε is the dielectric constant of the liquid; ε_0 is the permittivity of free space; ζ is the zeta-potential of the particles; E is the applied electric intensity; η is the viscosity of the liquid. If m_0 is the initial mass of the particles in the suspension, then:

$$m = m_0 - VC \tag{3.3}$$

where V is the volume of the suspension. Combining Eq. (3.1) and Eq (3.2) subject to the boundary condition of Eq. (6.3), we get

$$m = m_0(1 - e^{-\alpha t}) \tag{3.4}$$

$$\frac{dm}{dt} = m_0 \alpha e^{-\alpha t} \tag{3.5}$$

where $\alpha = Av/V$. According to Eq. (3.5), the rate of deposition decreases exponentially with time and is controlled by the parameter α.

3.4 Plastic Forming

Plastic deformation of a moldable powder-additive mixture is employed in several forming methods for ceramics[2]. Extrusion of a moist clay-water mixture is used extensively in the traditional ceramics sector for forming components with a regular cross section (e.g., solid and hollow cylinders, tiles, and bricks). The method is also used to form some oxide ceramics for advanced applications (e.g., catalyst supports, capacitor tubes and electrical insulators). A recent development is the repeated co-extrusion of a particle-filled thermoplastic polymer to produce textured microstructures or fine-scale structures. Injection molding of a ceramic-polymer mixture is a potentially useful method for the mass production of small ceramic articles with complex shapes. However, the method has not yet materialized into a significant forming process for ceramics mainly because of two factors:

(1) High tooling costs relative to other common forming methods.

(2) Removal of the high concentration of binder prior to sintering remains a limiting step for thicknesses greater than about 1 cm.

The application of extrusion and injection molding to the forming of ceramic powders has benefited considerably from the principles and technology developed in the plastics industry. Extruders and molding machines used in the plastics industry are employed but some modification of the machines is required for ceramic systems (e. g. , hardening of the contact surfaces). Two basic requirements must be satisfied for plastic forming to be successful:

(1) The mixture must flow plastically (above a certain yield stress) for the formation of the desired shape.

(2) The shaped article must be strong enough to resist deformation under the force of gravity or under stresses associated with handling.

The selection of additives and the formulation of the mixture are critical steps in meeting these requirements.

3.4.1 Extrusion

A slurry has its water content squeezed out in a filter press to produce a cake, which is then inserted into a vacuum pug mill, to mix and de-air, to get a mixing pug. The pug mill is used often to prepare a mix for extrusion as preliminary blending, its function of the pug mill is to homogenize the material by shear and to provide a more uniform batch to the extruder[2].

In extrusion, the mixing pug is compacted and shaped by forcing it through a nozzle in a piston extruder, as show in Figure 3.21, or a screw-fed extruder. The piston extruder is correspondingly simple in design, consisting of a barrel, a piston, and a die. In contrast, as illustrated by Figure 3.22, the screw extruder is complex, and considerable attention goes into the design of the extruder barrel and screw. The screw has to mix the powder and other additives into a homogeneous mass and generate enough pressure to transport the mixture against the resistance of the die. Shaping of the extruded body is

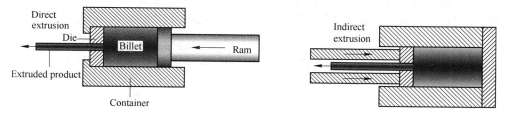

Figure 3.21 Schematic drawing of piston extruder

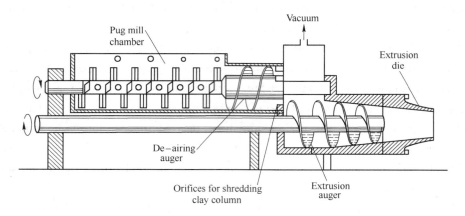

Figure 3.22 Schematic drawing of screw-fed extruder

achieved with the head of the extruder screw and the die. The extruder screw head changes the rotational flow of the mixture produced by the screw into an axial flow for extrusion and to produce uniform flow in the die. In the release of the body from the extruder, the die must generate the required cross section, allow uniform flow across the entire cross section, and ensure a smooth surface.

Figure 3.23 shows photos of piston extruder and screw-fed extruder for producing of SiC ceramic tube in Shandong Baona New Materials Co.

Figure 3.23 Photos of piston extruder (a) and screw-fed extruder (b)

The main approaches for imparting the required plastic properties to the feed material are: (1) manipulating the characteristics of the powder-water system, commonly used for clays; (2) adding a binder solution to the powder, commonly used for advanced ceramics. Clay particles develop desirable plastic characteristics when mixed with a controlled amount of water (15% ~ 30% (mass fraction) depending on the type of clay). The plasticity arises from two main factors: (1) particle to particle bonding due to the charged particle surfaces and intervening charges; (2) surface tension effects due to the presence of water. Surface charge and surface tension do not play a significant role in coarse ceramic powders, but some degree of plasticity may be developed by adding fine particles (such as clay or boehmite) provided that they are chemically compatible with the coarse powder. The powders of advanced ceramics, when mixed with water, do not possess the desirable plastic characteristics found in the clay-water system. For this reason, they are mixed with a viscous solution containing a few weight percent of an organic binder to provide the desired plastic characteristics. The solvent is commonly water, but non-aqueous solvents (e.g., alcohols; mineral spirits) can also be used. Since the extruded body must also have sufficient green strength, the binder is generally selected from the medium to high viscosity grades, e.g., methylcellulose, hydroxyethyl cellulose, poly (acrylimides), or poly (vinyl alcohol). Methylcellulose undergoes thermal gelation, a property that offers considerable benefits for extrusion of ceramics.

Examples of compositions used in extrusion are given in Table 3.6. Aqueous systems are often flocculated using a small concentration of an additive (such as $MgCl_2$, $AlCl_3$, or $MgSO_4$). In addition, lubricants (e.g., stearates, silicones, or petroleum oils) are commonly used to reduce die-wall friction.

Table 3.6 Examples of compositions used in extrusion[2]

Whiteware		Alumina	
Material	Concentration (volume fraction) /%	Material	Concentration (volume fraction) /%
Kaolin	16	Alumina	45 ~ 50
Ball clay	16	Water	40 ~ 45
Quartz	16	Ammonium polyacrylate (dispersant)	1 ~ 2
Feldspar	16	Methyl cellulose (binder)	5
Water	36	Glycerin (plasticizer)	1
$CaCl_2$ (flocculant)	<1	Ammonium stearate (lubricant)	1

The organic materials are frequently used in extrusion of advanced ceramics to provide plastic flow. Not only plastic behaviour is important for the extrusion of ceramic bodies. There are many other characteristics that can be tailored by the suitable addition of organics in a ceramic extrusion paste, or feedstock.

A method based on repeated co-extrusion of a powder-filled thermoplastic polymer has been developed recently to form ceramics with a textured microstructure or with fine-scale features. This process is known as **thermoplastic extrusion of ceramic bodies**.

Thermoplastic materials are polymers which, when heated, soften, melt or become more pliable, and harden during cooling in a reversible physical process. Materials in this class which are used quite often for ceramic processing are PE, PP, EVA, POM and PMMA. One of the main advantages of thermoplastic systems for ceramic extrusion is the lower abrasivity of the feedstock material relative to other binder systems. Another advantage of thermoplastic binder systems over solvent-based ones is the contour accuracy of the extruded material, which permits easy fabrication of fine structures (e. g. , thin-walled tubes, micro-tools).

Generally for the thermoplastic extrusion of ceramic bodies, a binder which melts or softens at higher temperature is used. Figure 3.24 shows the processing steps for the extrusion of ceramic bodies with thermoplastic binder systems. The figure shows the process schematically starting from the raw material components through to the finished ceramic product.

It is also possible to extrude ceramic nanopowders with thermoplastic binder systems. Scheying et al, for example, investigated the extrusion of monoclinic zirconia nanopowder[12]. The powder loadings achieved in the feedstocks were 44% and 53% (volume fraction) when using powders with primary particle sizes of 9 and 25nm, respectively. The binder system used consisted of either polyethylene-co-vinyl acetate or polyethylene as the major component and decanoic acid as a surfactant.

A variety of defects can occur in the extruded ceramic body. The common **macroscopic defects** are laminations, tearing, and segregation. Laminations are cracks that generally form a pattern or orientation, particularly in a screw fed extruder, because of incomplete re-knitting of the feed material around the auger. Tearing consists of surface cracks that form as the material leaves the die and is caused by poor die design or by low plasticity of the mixture. Segregation involves separation of the liquid and the solid phases of the mixture during extrusion and is often caused by poor mixing. Microscopic defects such as pores (caused by trapped air) and inclusions (due to contamination) can also occur.

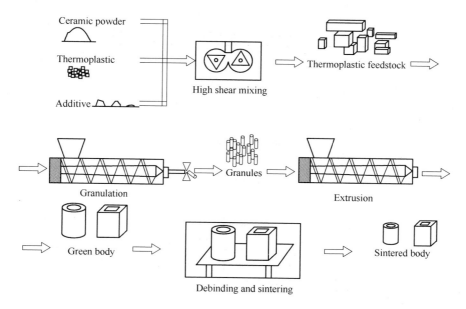

Figure 3.24 Process schematic for thermoplastic extrusion of ceramic bodies[16]

3.4.2 Injection

Ceramic injection molding (CIM) has become a standard process in the ceramics industry in the last 20 years. The injection molding is similar to thermoplastic extrusion of ceramic bodies, it is another technique that is widely used in shaping thermoplastic polymers. A thermoplastic polymer is one that softens when heated and hardens when cooled. Such processes are totally reversible and may be forming ceramic components if the ceramic powder is added to a thermoplastic polymer. The production of a ceramic article by injection molding involves the following steps: selection of the powder and the binder; mixing the powder with the binder, production of a homogeneous feed material in the form of granules, injection molding of the green body, removal of the binder (debinding) at lower temperatures and, finally, sintering at higher temperatures to produce a dense, final article.

When forming ceramics by injection molding, the polymer is usually referred to as the binder. The ceramic powder is added to the binder and is usually mixed with several other organic materials to provide a mass that has the desired rheological properties. Table 3.7 shows the additives that have been used to form SiC shapes by injection molding. The organic part of the mix accounts for about 40% (volume fraction).

Table 3.7 Additives for injection forming of SiC[1]

Function	Example	Quantity (mass fraction)/%	Volatilization temperature/℃
Thermoplastic resin	Ethyl cellulose, polyethylene, polyethylene glycol	9~17	200~400
Wax or high-temperature volatilizing oil	Paraffin, mineral oils, vegetable oils	2~3.5	150~190
Low-temperature volatilization Hydrocarbon or oil	Animal oils, vegetable oils, mineral oils	4.5~8.5	50~150
Lubricant or mold reiease	Fatty acids, fatty alcohols, fatty esters	1~3	
Thermosetting resin	Epoxy, polyphenylene, phenol formaldehyde		Gives carbon 450~1000

The plastic mass is first heated, at which point the thermoplastic polymer becomes soft and is then forced into a mold cavity as shown in Figure 3.25. The heated mixture is very fluid and is not self-supporting (this is different from the situation encountered in extrusion). The mixture is allowed to cool in the mold during which time the thermoplastic polymer hardens. Because of the large volume fraction of organic material used in the mixture, there is a high degree of shrinkage of injection-molded components during sintering. Shrinkage of 15%~20% is typical, so precise control of component dimensions is difficult. However, complex shapes are retained with very little distortion

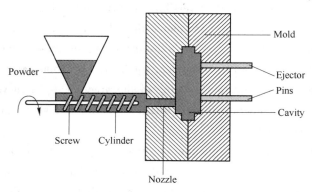

Figure 3.25 Schematic drawing of injection molding

during sintering since the densities, although low, are uniform.

Figure 3.26 and Figure 3.27 are the photos of a granulator and a vertical injection molding machine using in our laboratory. With these two machines, we have successfully made the bodies of pressureless sintering SiC with solid loading 70% (mass fraction).

Figure 3.26 Photo of granulator

Figure 3.27 Vertical injection molding machine

During ceramic injection molding, the presence of organic binder is to facilitate flowability of the powder-binder suspension into mould cavity and to keep the particles attached firmly. The organic binder is temporary and will eventually be removed in subsequent thermal cycle. The removal of organic binder is a procedure (usually termed debinding) that is known to be critical in obtaining a green compact with desired microstructure[17]. Here we emphasize the debinding of organic binders.

Prior to sintering, the binder system must be completely removed from the extruded parts. Compared to solvent-based binder systems, thermoplastic binders are more difficult to remove from extruded bodies. In solvent-based systems, the removal of solvent during the drying step immediately after extrusion results in open porosity being formed through which the decomposition products of the high-molecular weight polymers can be easily transported during subsequent thermal binding treatment. In thermoplastic debinding, this initial open porosity does not exist, and poorly controlled decomposition of the binder can easily lead to bubble and crack formation due to the build-up of internal pressure. In the literature several different binder removal processes are described[11,18~20].

In principle thermoplastic binder systems can be removed from ceramic green bodies by thermal heat treatment. Unlike in solvent-based binder systems, where some initial porosity is available after removal of the solvent (drying step), there is no open porosity in thermoplastic green extrudates at the beginning of the debinding process. To ensure complete burnout of the binder, diffusion of oxygen into the green body and diffusion of the evaporated, decomposed or combusted organic products out of the body must be guaranteed. Waxes (short-chain polymers) are removed in the gaseous state and porosity is created during the first step of the debinding process of the thermoplastic binder system. Subsequently, the decomposition products of the major binder component can escape through the created porosity without causing any defects in the ceramic bodies prior to sintering.

Calvert et al describes a theoretical model to calculate the burn-out time as a function of the extrudate thickness without "bubble" formation[21]. By increasing early development of porosity within the sample, higher debinding temperatures can be used and consequently faster burn-out can be achieved. Furthermore the products of the binder decomposition should have a high vapour pressure and should diffuse rapidly through the extruded body. For bodies with thick walls, the initial creation of open porosity is critical. During heat treatment the evaporation and decomposition of the polymers generate gas inside the body. If the pores are not open or there are not enough paths to channel the gas to the surface of the extrudate, pressure will build up inside the body and cracks will be created during the debinding process. Therefore binder components with different decomposition characteristics (e. g. melting points, partial vapour pressures) are used. Thermal binder removal is very sensitive to changes in heating rates, dwell times and debinding atmospheres, and therefore strict process control is indispensable.

3.4.2.1 Binder removal by wicking

Another option is to build up an open pore structure in the ceramic body by wicking out parts of the binder system based on the capillary effect. Generally binder removal can be described with three mechanisms: fluid wicking, diffusion control, and permeation control. In this approach the green ceramic body is placed on a porous substrate or embedded in a fine ceramic powder. When the temperature of the ceramic body exceeds the melting temperature of the lowest melting component, capillary forces draw (wick) the resulting fluid components out of the ceramic body into the surrounding porous material. To ensure that this wicking process takes place, the average pore size of the powder bed should be lower than that found in the green ceramic body. After the melted components have been wicked into the powder bed and some initial open porosity has been created, normal thermal debinding (diffusion and permeation control) can be used to remove the rest of the binder components in the body.

3.4.2.2 Binder removal by solvent extraction

To create an initial porous structure before thermal binder removal, it is also possible to extract parts of the binder system by solvent extraction. In this technique the ceramic green bodies are immersed in a liquid which solvates one of the components in the binder used. A prerequisite of this process is that the binder component to be removed first has a high solubility in the solvent to ensure rapid dissolution. Furthermore the viscosity of the resulting polymer solution should be low, and diffusion of the liquid inside the body should be fast. Swelling of the binder components during solvent extraction can cause defects to be formed inside the ceramic body because of the internal stresses generated. With the solvent extraction process, the speed of binder removal is a relatively fast and the cost of the equipment relatively low.

3.4.2.3 Binder removal by supercritical extraction

Supercritical extraction involves a gas under supercritical conditions where it changes into a supercritical fluid. The diffusion coefficient of the supercritical fluid is similar to that of the gas, and the solubility is similar to the liquid phase. The big advantage is that such a fluid has no surface tension, and therefore no capillary forces are developed by the fluid in the body. Consequently, inhomogeneous regions in the body (e.g., agglomerates, micro-defects and density variations) are not so critical. Various gases can be used for this removal step, however, carbon dioxide is the most common because it is safe and cost-effective.

3.4.2.4 Binder removal by catalytic process

By using a catalyst the polymeric components at the surface of the green body are cracked into monomers and evaporate. The porosity thus created exposes virgin polymer beneath the surface and the catalytic process continues deeper into the sample, debinding the green body from the outside inwards. With this technique, short debinding times can be achieved. The big advantage of this process is that binder removal takes place at temperatures lower than the melting temperature of the polymers, and consequently warping during the debinding process is not such a problem.

Advanced ceramics with submicron or even nanoscale structural features are well known from material science. Due to the difficult processing properties of such ultrafine powders, nanocomposites produced by injection molding are not yet commercially available. Many nanocomposite ceramics with extremely attractive properties as extreme hardness, toughness and strength are only applied in high-price niche applications due to their high manufacturing cost which is determined by low turnover hot-pressing or spark plasma sintering (SPS) processes and difficult and cost intensive hard machining operations. F. Kern and R. Gadow[22] reported extrusion and injection molding of ceramic micro and nanocomposites. Their research results showed that the main difficulties identified in nanocomposite feedstocks were the very narrow ranges of composition and processing parameters which have to be exactly complied to in order to avoid binder segregation, warping and microstructural defects in the sintered part.

Rudolf Zauner[23] introduced micro powder injection molding. His results showed that the smallest achievable structures were about ten times the particle size. For example, with 400nm alumina particles, structures of 5 μm were successfully injection molded and sintered. As a wide variety of metals and ceramics can be processed, including functional and catalytic materials, micro powder injection molding opens up new avenues for the mass production of microcomponents for the medical and automotive industries and for applications in (bio) chemistry and sensor technology.

3.4.3 Roll Forming

A plastic mix is passed between two cylinders that are rotating in opposite directions as shown in Figure 3.28. The plastic mix passing between the rolls is compacted, as well as being pressed to a thickness equivalent to the spacing of the rolls. Multiple passes at diminishing roll separation can yield a constant thickness sheet of high uniformity.

Roll forming can be conducted at room temperature using a mix equivalent to an extrusion mix or at elevated temperature using a thermoplastic polymer system. Warm roll forming has been used for many years to fabricate resin-bonded and rubber-bonded grinding wheels. It has also been used to form much thinner layers suitable for heat-exchanger fabrication.

Recently J. R. G. Evans[24] published a paper that introduced seventy ways to make ceramics. In fact, each ceramic forming method is suited for some applications and unsuited for others. Thus, in production planning, it is necessary to choose the appropriate forming method after careful consideration of the shape of the object to be produced,

3.4 Plastic Forming · 131 ·

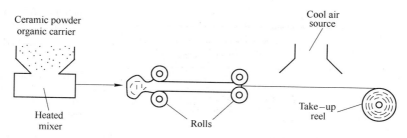

Figure 3.28 Schematic illustrating the roll-forming process[3]

but also the delivery time, the quantity needed, the cost, and other factors. The amount and type of binder used are also large decided by the forming method. Thus, the ultimate decision about the molding technique used requires consideration of the characteristics of the ceramic body. Table 3.8 summarizes the distinctive features of the major methods of forming ceramics.

Table 3.8 Types of molding methods and their characteristics[25]

Molding methods	Molding additives (amount per 100 parts of ceramic powder)	Shape of products
Press forming	Binder: Water-soluble resin Emulsion (3.0-5.0) Gum Arabic, etc Water (0.3-1.0)	Small, flat objects Various containers for electronic components Liners for mills
Isostatic pressing	Binder: Water-soluble resin Emulsion (2.0-5.0) Gum Arabic, etc Water (0.3-1.0)	Tubular, columnar, or spherical forms Balls for grinding Insulators for spark plugs Outer tube for vacuum tubes
Injection molding	Binder: Thermoplastic resin Wax, etc (10.0-25.0) Plasticizer Phthalic ester Various oils, etc (0.5-5.0)	Mass production of products in complex shapes Thread guides for spinning Valves Turbine vanes
Doctor tape method	Binder Polyvinyl butyl Acrylic ester, etc (8.0-15.0) Plasticizer Phthalic ester, etc (3.0-8.0) Solvent Alcohols, ketones, Hydrocarbons, etc (5.0)	Thin sheets (1.5mm or less thick and secondary processed products of such sheets Laminated packages Multilayer substrates Thin-and thick-film substrates

Continues Table 3.8

Molding methods	Molding additives (amount per 100 parts of ceramic powder)	Shape of products

Slip casting	Binder
	  Sodium alginate
	  Methyl cellulose, etc, } (0.5-3.0)
	Water (15.0-30.0)
		Thin-walled products in irregular Shapes
		Crucibles
		Thread guides
		Blades

Extrusion	Binder
	  Wax emulsion
	  Water-soluble resin } (8.0-15.0)
	Water (15.0-30.0)
		Long objects in columnar, tubular, etc, forms
		Insulator tubes
		Protective tubes
		Ceramic honeycomb
		Pipes

Hot-press molding	Normally no additives are used	Especially objects which require High density and high strength Ceramic tools Piezoelectric substances

Technical point	Advantages	Disadvantages
Adjustment and control of granules Pressing conditions Pressure Pressing speed Duration of pressing, etc	Least expensive method of forming Readily automated Product can be fired directly after molding	Density tends to be uneven in forming, limiting the sizes and shapes which can be molded

Adjustment and control of granules
Pressing conditions
  Selecting rubber quality and thickness
  Receptivity to cutting and finishing processing
  Reuse of shavings
	Density of the formed objects is uniform with few distortions of shape
	  The dry method permits mass production
		Many cases require external shaping and cutting after forming
		  The life of the rubber mold is short
		  The wet method is not commercially practicable

Kneading the body and binders Dewaxing Injection conditions Designing the mold	Can mass-produce products in complex, three-dimensional shapes Outstanding dimensional accuracy Surface finish is good	Mold costs are high, making the method unsuited to small-scale production The dewaxing process is difficult

Slurry control
  Preventing stretching or shrinking of the sheets
  Selection of the forming aid to give the material sheet characteristics such as the ability to be laminated
	Outstanding productivity and good accuracy in the thickness produced
	A range of secondary processing of the sheets is possible
	No dimensionality in the shrinkage in firing
		The equipment is expensive
		  The equipment takes up a great deal of shape
		  Care must be given to explosion prevention measures and to health and safety

Continues Table 3.8

Technical point	Advantages	Disadvantages
Slutty control Design of the plaster-of paris mold The drying method	Can be produced with simple equipment Can creat relatively complex shapes	Distortions frequently occar in the casting The casting process is time-consuming Many plaster of paris molds are needed, taking up space
Design of the dies Dryng method	No special limitation on the length of the object to be extruded Continuous production is possible	A drying step is needed after the forming When objects with large cross sections are to be formed, the equipment is large
The equipment as a whole Choosing the appropriate pressure	The density after sintering is increased, and physical properties are improved It is possible to reduce the sintering temperature	There are major limitation on forms Apart from a few special products, lacks suitability for mass production

Study Guide

3-1   What are the purposes of the additives to powder that is to be used to form a ceramic part by pressing?

3-2   "Dry pressing" is used extensively for low cost production of large quantities of simple ceramic parts in automated presses. Identify typical parameters for dry pressing, and explain what happens to the powder during dry pressing.

3-3   What are some potential causes of cracking during uniaxial pressing?

3-4   Explain potential sources of nonuniform density during uniaxial pressing.

3-5   What is "isostatic pressing" and what are the benefits?

3-6   Compare wet-bag and dry-bag isostatic pressing.

3-7   What are some important ceramic products manufactured by dry-bag isostatic pressing?

3-8   Briefly describe how a ceramic part is made using slip casting.

3-9   What is a dispersant, and why is it important for slip casting?

3-10   What is the most common mold material for slip casting?

3-11   Casting can be done in a variety of ways. List some of these options.

3-12   How does "gel casting" differ from other forms of casting a slip?

3-13   What is tape casting, and how does it differ from slip casting?

3-14   What are some important applications of tape casting?

3-15   What is extrusion, and what are the shape limitations?

3 – 16 Identify some important ceramic products fabricated by extrusion.

3 – 17 How does injection molding differ from extrusion?

3 – 18 Binder removal is a very critical step in forming a ceramic by injection molding. Discuss various mechanisms of binder removal.

3 – 19 What is "green machining"?

References

[1] Carter Barry C, Norton Grant M. Ceramic Materials, Science and Engineering [M]. Springer Science + Business media, LLC, 2007.

[2] Rahaman M N. Ceramic Processing and Sintering [M]. Marcel Dekken Inc., 2003.

[3] David W Richerson. Modern Ceramic Engineering [M]. Third Edition. Taylor & Francis Group, 2006.

[4] Rowlands, R R. A Review of the Slip Casting Process [J]. Am. Ceram. Soc. Bull., 1966, 45 (1): 16~19.

[5] Rado P. Slip-casting of Non-clay Ceramics [J]. Interceram, 1987, 36 (4): 34~37.

[6] Rempes P E, Weber B C, Schwartz M A. Slip Casting of Metals, Ceramics, and Cermets [J]. Am. Ceram. Soc. Bull., 1958, 37 (7): 334~339.

[7] Smith P A, Kerch H M, Haerle A G, et al. Microstructural Characterization of Alumina and Silicon Carbide Slip-Cast Cakes [J]. Am. Ceram. Soc., 1996, 79 (10): 2515~2526.

[8] Williams R M, Ezis A. Slip Casting of Silicon Shapes and their Nitriding [J]. Am. Ceram. Soc. Bull., 1983, 62 (5): 607~610, 619.

[9] Hoffmann M J, Nagel A, Greil P, et al. Slip Casting of SiC-Whisker-Reinforced Si_3N_4 [J]. Am. Ceram. Soc., 1989, 72 (5): 765~769.

[10] Suzuki S, Nasu T, Hayama S, et al. Mechanical and Thermal Properties of β-sialon Prepared by a Slip Casting Method [J]. Am. Ceram. Soc., 1996, 79 (6): 1685~1688.

[11] Schmitt C, Agar D W, Platte F, et al. Ceramic Plate Heat Exchanger for Heterogeneous Gas Phase Reactions [J]. Chemical Engineering & Technology, 2005, 28 (3): 337~343.

[12] Tok A I Y, Boey F Y C, Khor K A. Tape Casting of High Dielectric Ceramic Composite Substrates for Microelectronics Application [J]. Journal of Materials Processing Technology, 1999, 89: 508~512.

[13] Omatete O O, Janney M A, Nunn S D. Gel casting: from Laboratory Development toward Industrial Production [J]. Journal of the European Ceramic Society, 1997, 17 (2): 407~413.

[14] Corni I, Ryann M P, Boccaccini A R. Electrophoretic Deposition: from Traditional Ceramics to Nanotechnology [J]. Journal of the European Ceramic Society, 2008, 28 (7): 1353~1367.

[15] Scheying G, Wuhrl I, Eisele U, et al. Monoclinic Zirconia Bodies by Thermoplastic Ceramic Extrusion [J]. Am. Ceram. Soc, 2004, 87 (3): 358~364.

[16] Clemens Frank. Thermoplastic Extrusion for Ceramic Bodies [M] //Extrusion in Ceramics. Springer Berlin Heidelberg, 2007: 295~311.

[17] Liu Dean-MO, Tseng Wenjea J. Influence of Debinding rate, Solid loading and Binder Formulation on the Green Microstructure and Sintering Behaviour of Ceramic Injection Moldings [J]. Ceramics International, 1998, 24 (6): 471~481.

[18] Trunec M, Cihlar J. Thermal Removal of Multicomponent Binder from Ceramic Injection Mouldings [J]. Journal of the European Ceramic Society, 2002, 22 (13): 2231~2241.

[19] Onbattuvelli V P, Enneti R K, Park S J. The Effects of Nanoparticle Addition on Binder Removal from Injection Molded Aluminum Nitride [J]. International Journal of Refractory Metals and Hard Materials, 2013, 36: 77~84.

[20] Xie Z, Huang Y, Wu J, et al. Microwave Debinding of a Ceramic Injection Moulded body [J]. Journal of Materials Science Letters, 1995, 14 (11): 794~795.

[21] Calvert P, Cirma M. Theoretical Models for Binder Burnout [J]. Am. Ceram. Soc. 1990, 73 (3): 575~579.

[22] Kern F, Gadow R. Extrusion and Injection Molding of Ceramic Micro and Nanocomposites [J]. International Journal of Material Forming, 2009, 2 (1): 609~612.

[23] Zauner R. Micro Powder Injection Moulding [J]. Microelectronic Engineering, 2006, 83 (4): 1442~1444.

[24] Evans J R G. Seventy Ways to Make Ceramics [J]. Journal of the European Ceramic Society, 2008, 28 (7): 1421~1432.

[25] Shigeyuki Somiya. Advanced Technical Ceramics [M]. Academic Press Japan Inc., 1984.

4 Sintering

The process of heat treatment of ceramic body is described by two terms "firing" and "sintering".

After the binder has been removed from the molded or cast object (still an aggregate of ceramic powder), it is fired and hardened into a new state, becoming a unified, dense object. This process is called firing. Firing, the process of heat treatment of ceramic ware in a kiln to develop a vitreous or crystalline bond, thus giving the ware properties associated with a ceramic material.

About "Sintering", many authors have given different definition. **Sintering** is a processing technique used to produce density-controlled materials and components from metal or/and ceramic powders by applying thermal energy[1]. **Sintering** is the process of transforming a powder into a solid body using heat. The **densification** of a particulate ceramic compact is technically referred to as **sintering**[2]. When thermal energy is applied to a powder compact, the compact is **densified** and the **average grain size increases**. The basic phenomena occurring during this process, called **sintering**, are densification and grain growth[3]. Sintering is essentially a removal of the pores between the starting particles accompanied by shrinkage of the component, combined with growth together and strong bonding between adjacent particles. Sintering, the general term for the densification may be by the formation of liquid phases which fill pores; by solid state diffusion or other mechanisms which also reduce surface energy. In a word, sintering is the densification process of a particulate ceramic compact.

In the field of advanced ceramics, typically in structural ceramics, the products often need to be densified. Therefore, in this chapter, we focus on the sintering processing of ceramics, and the sintering theory will be discussed in Material Science course.

4.1 Kiln and Furnace

In this chapter, the equipment to heat treatment of ceramic ware is called as furnaces, but it could have been called kilns or ovens, as all these terms are used to describe many of the same types of equipment. "Kiln" is widely used in the traditional ceramics industry. "Furnace" is used interchangeably with kiln. "Oven" is more often used for either equipment used for drying ceramics (typically using lower temperatures) or for small furnaces.

Kiln is a high-temperature installation used for firing ceramic ware or for calcining or sintering[4]. Kilns for firing ceramic ware are three main types: intermittent (batch), annular and tunnel; for calcining and sintering: shaft kilns, rotary kilns and multiple-hearth furnaces are used.

A furnace is a device that produces heat. Not only are furnaces used in the home for warmth, they are used in industry for a variety of purposes such as making steel and heat treating of materials to change their molecular structure.

In practice, it is difficult to make a distinction between kiln and furnace when we refer the firing equipment.

4.1.1 Classification of Ceramic Furnaces (Kilns)

Furnaces are the essential equipment in any ceramics laboratory and factory. They can range in size from small electrically heated box furnaces, which can fit on a bench, to the enormous gas-fired furnaces used to produce quartz ceramic crucible. In between these extremes there are furnaces of many shapes and sizes, designed to run at a range of temperatures and in a range of atmospheres. In addition to obtaining a high temperature, it is necessary to have furnace components that can withstand these temperatures without degradation. These materials are known as refractories. Whenever temperatures are high, vapor pressures may also be high.

Based on the mode of charging of material, furnaces can be classified as batch type furnace and continuous furnace. Based on the method of generating heat, furnaces are broadly classified into two types namely combustion type (using fuels) and electric type.

In batch kilns, the objects to be fired are loaded in the kiln, fired, cooled, and removed; this set of operations is carried out on a single-batch basis. Thus, batch kilns are not well suited for large-scale mass production, but they are the appropriate choice for processing especially large or long objects. The batch kiln also has the advantages of being flexible in regard to firing conditions and of being relatively inexpensive to build.

The continuous kiln is used for firing high-volume products. Figure 4.1 illustrates the tunnel-shaped continuous kiln. It has three zones, for preheating, firing, and cooling. The products are loaded on carts, which enter the kiln from one end, one by one, at a given interval of time. The cart advances on the rails a certain distance, then, according to the temperature distribution created within the kiln, goes through the preheating, firing, and cooling processes, and is removed at the opposite end. In a continuous kiln, the temperature at each location is constant with time. The parts are moved through the furnace at a velocity giving the desired time-temperature profile. Continuous kilns are best for mass production where large quantities of material are subjected to the same conditions. The disadvantages of continuous kilns are that the kiln temperature must be maintained throughout the process and their lack of flexibility.

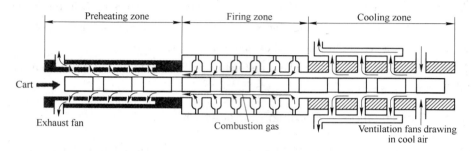

Figure 4.1 Structure of a tunnel-shaped continuous kiln

In the case of combustion type, depending upon the kind of combustion, the kilns can be broadly classified as oil fired, coal fired or gas fired. But the most common combustion kilns used in ceramic processing are gas fired and use gaseous hydrocarbons as fuel. The large amount of energy produced heats the kiln and the parts inside. Gas-fired kilns are used mainly in large industrial applications such as the production of alumina ceramics, and firing of traditional ceramic articles, for example, white-ware.

Combustion is an oxidative process. Table 4.1 lists the standard molar enthalpies of combustion for a range of hydrocarbons. The amount of energy released during com-

Table 4.1 Standard molar enthalpies of combustion, $H_c^{\ominus}$ [1]

Hydrocarbons	Standard molar enthalpies $\Delta H_c^{\ominus}/kJ \cdot mol^{-1}$	Hydrocarbons	Standard molar enthalpies $\Delta H_c^{\ominus}/kJ \cdot mol^{-1}$
$CH_4(g)$	890	$C_{10}H_8(s)$	5157
$C_2H_2(g)$	1300	$CH_3OH(l)$	726
$C_2H_4(g)$	1411	$CH_3CHO(g)$	1193
$C_2H_6(g)$	1560	$CH_3CH_2OH(l)$	1368
$C_4H_{10}(g)$	2877	$CH_3COOH(l)$	874
$C_6H_{12}(l)$	3920	$CH_3COOC_2H_5(l)$	2231
$C_6H_{14}(l)$	4163	$C_6H_5OH(s)$	3054
$C_6H_6(l)$	3268		

bustion depends on the strength of the bond in the gas. Fuels with many weak (less stable) bonds, such as C—C and C—H, yield more energy than fuels with fewer such bonds or fuels that contain large numbers of strong bonds, e. g. , C—O and O—H.

The maximum temperature that can be achieved in a gas-fired kiln is well below the adiabatic flame temperature because of heat loss caused by incomplete combustion and dissociation of the combustion gases (an endothermic process) at high temperature. Heat is also lost by conduction through the refractories and imperfect insulation.

Electrical furnaces can produce direct (resistive) or indirect (induction or microwave) heating. Most electrically heated furnaces use the principle of Joule, or resistance, heating where current flowing through a resistor produces heat. The heat source in the earliest electric furnaces was dc arcs formed between carbon electrodes, so the heating element really was an element. Carbon (in the form of graphite) is still used as a heating element, but most heating elements are now made from compounds. There are several different types of heating element and we will describe some of the important ones in this section. The choice of heating element depends on the maximum temperature that is required and the environment to which the element will be exposed.

The main advantage of electric heating is that it is easy to measure power input, to control heating rates and temperature, and that the furnace can operate in an atmosphere independent of the heating source.

The main disadvantage of electric heating is that it usually costs more per energy unit than gas heating. However, the total energy usage for electric furnaces may often be lower than for gas-fired furnaces. The cost issue is not normally a problem in a university laboratory, but it can be a major concern for industrial applications.

4. 1. 2 Batch kilns

4. 1. 2. 1 Small furnaces

Here we call these small furnaces, it is that these furnaces used widely in laboratories or in many small-sized industrial applications.

Figure 4. 2 shows muffle furnace using in our laboratory, which can be used at maximum 1350℃ with SiC rods as the heating elements. Figure 4. 3 shows a 1800℃ high temperature box furnace with $MoSi_2$ parts as heating elements.

Figure 4. 4 shows an example of a small electrically heated furnace. This type of furnace is used for temperatures up to 2300℃ and are designed to be used in vacuum or inert atmosphere in our laboratory. Figure 4. 5 shows another example of a tube furnace. This particular furnace is known as a tube furnace. By flowing gases along a tube

Figure 4.2　Box furnace

Figure 4.3　1800℃ High temperature box furnace

Figure 4.4　Electrically heated vacuum furnace

Figure 4.5　Electrically heated tube furnace

placed inside the furnace the heating environment can be controlled.

4.1.2.2　Hot press sintering furnace

Vacuum hot-pressing furnace are mainly used for nonmetal materials, carbon/carbon composite materials, ceramic materials and metal powder materials in vacuum or protecting atmosphere conditions sintering production and hot-press experiments. Figure 4.6 shows a hot press sintering furnace in our laboratory. The parameters of this furnace are: max temperature 2300℃, RT-2200℃ 60min, test – piece case $\phi120 \times 110H$, vacuum degree 6.65×10^{-3}Pa, max gas (N_2 or Ar) pressure 0.93 MPa, total pressure 4.9×10^4N.

4.1.2.3 Inductuin sintering furnace

Induction Heating provides a means for precise heating of electrically conducting objects. The object is immersed in an alternating magnetic field, which is usually produced by an external coil energized by an ac-source. The magnetic field induces voltages in the conductive material and these voltages produce circulating currents (called eddy currents). The magnitude of the induced voltage and the impedance of the material determine the size of the induced currents. It is the flow of induced currents that produces Joule heating of the material. If the material we want to heat is an insulator we can place it inside a conductive crucible, such as graphite. A typical induction furnace is shown in Figure 4.7.

Figure 4.6 Electrically heated hot press sintering furnace

Figure 4.7 An induction furnace using to produce reaction sintering SiC products

Induction heating gives following advantages: (1) It is clean and fast; (2) The process is easily reproducible; (3) It can be automated; (4) Localized heating is possible.

Induction furnaces operate at frequencies from 60 to 1000 Hz and are thus often referred to as RF furnaces. They can be used to obtain temperatures up to 3000℃. Since the coil currents may be as high as 15 kA, the Cu coil conductors are usually hollow to permit water circulation for cooling. Induction furnaces are generally used for melting and surface hardening. They are sometimes used in sintering in conjunction with hot pressing.

4.1.2.4 Microwave sintering furnace

Microwaves are electromagnetic waves with wavelengths ranging from 1m to 1mm, which correspond to frequencies between 0.3 and 300GHz[5]. This frequency range lies just above radio waves and just below visible light on the electromagnetic spectrum. 0.915GHz and 2.45GHz frequencies are commonly used for microwave heating. These frequencies are chosen for the microwave heating based on two reasons. The first is that they are in one of the industrial, scientific and medical (ISM) radio bands set aside for non-communication purposes. The second is that the penetration depth of the microwaves is greater for these low frequencies. However, heating is not necessarily increased with decreasing frequency as the internal field can be low depending on the properties of the material. 2.45GHz is mostly used for household microwave ovens and 0.915GHz is preferred for industrial/commercial microwave ovens. Recently, microwave furnaces with variable frequencies from 0.9 to 18GHz have been developed for material processing. Microwaves are coherent and polarized and can be transmitted, absorbed, or reflected depending on the material type[6].

Microwave heating is an application of induction heating using higher frequencies. Heat is generated in non-conducting materials when microwave radiation excites the molecules in the material. The high-frequency radiation causes molecular polarization and the ability of the molecules to follow the rapid reversal of the electric field results in the conversion of electromagnetic energy into heat within the irradiated material.

Microwave furnace consists of three major components: the source, the transmission lines and the applicator.

(1) Microwave generator. Microwave generators consist of a magnetron for producing microwave energy. Magnetrons are made up of a cathode enclosed in circular anode cavities. These cavities are surrounded by powerful magnets designed for producing strong magnetic fields inside the cavity. The cathode produces electrons when heated (using high currents), these are accelerated towards anode. Due to the surrounding magnetic field the electrons move in a spiral path inducing alternating currents in the anode cavities. This will result in the generation of electromagnetic wave. It is the size of the anode cavity that determines the frequency of the electromagnetic radiation. The output of this field is transfered into the transmission line (wave guide) using a small loop. The transmission line would essentially transfer the electromagnetic radiation to the applicator containing load (material to be heated).

(2) Transmission line. Waveguides provide a viable means of transmitting high power microwaves, from generator to the load. These are essentially hollow metallic tubes either rectangular or circular in shape, made using brass, copper or aluminum.

The main objective of the transmission line is to provide an efficient means of transferring microwave energy from generator to the applicator.

(3) Applicator. Applicators are essentially hollow metallic cavities that contain load (in some cases are similar to waveguides). The microwave signal fed into these cavities will suffer multiple reflections along the preferred directions. The forward and reflected waves in the cavity superimpose on one another, producing a standing wave pattern.

There are many different types of microwave cavities. These are mainly classified into singlemode and multimode cavities. Singlemode cavities allow only a single wave pattern. Cavities that have more than one type of wave pattern are referred to as multimode applicators. Multimode applicators are designed to support large number of resonant modes at a given frequency. The advantage of multimode applicators is that they are versatile in accepting a wide range of heating loads. The advantage with single mode cavities is the well-defined field distribution that can be measured precisely. The type of applicator used depends on the materials to be processed. The single mode applicator and the traveling wave applicator are successful in processing materials of simple geometries. However, the multimode applicator has the capability to produce large and complex components. Therefore, multimode systems are used for industrial applications[5].

Figure 4.8 shows the sample location, the susceptor design and the insulation of a microwave hybrid (combination of microwave and susceptor heating) sintering system. The susceptor❶ shown in Figure 4.8b was made of 2% (mass fraction) partially stabilized ZrO_2 with 98% (mass fraction) Al_2O_3 composite. This particular composition allowed the susceptor to supply heat to the sample in a conventional manner.

4.1.2.5 Spark plasma sintering furnace

Spark Plasma Sintering (SPS) also known as Field Assisted Sintering Technology ("FAST"), SPS-furnaces are partly similar to conventional hot presses. Figure 4.9 shows a photo of SPS furnace and Figure 4.10 shows the principle scheme of SPS. Likewise there is a hydraulic pressing system, a water cooled vacuum chamber, a gas/vacuum control system, but—the main difference—a very special power supply system as well as a special tool design. This enables electrical current flowing directly through the sample and/or the die, depending on the electrical conductivity of the components, making very high heating and cooling rates possible with a relatively low energy consumption. Beneath the benefit of direct Joule heating the use of high current DC pulses

---

❶ A susceptor is a material used for its ability to absorb electromagnetic energy and convert it to heat. This energy is typically radiofrequency or microwave radiation used in industrial heating processes.

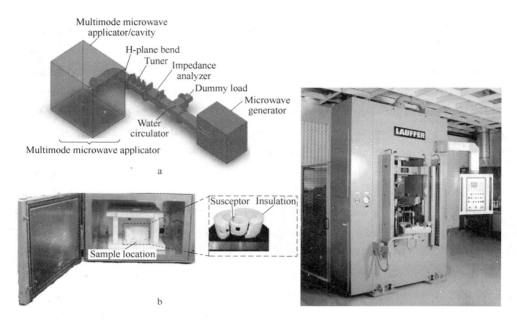

Figure 4.8 Schematic sketch of the multimode microwave system used for performing hybrid sintering experiments (a) and Multimode microwave cavity showing the insulation, the susceptor design and the sample location (b)[7]

Figure 4.9 Photo of a HP D 100 spark plasma sintering furnace (From FCT Systeme GmbH)

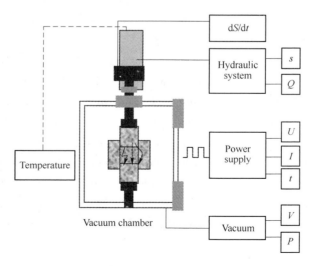

Figure 4.10 Principle scheme of Spark Plasma Sintering

can give an additional advantage by the increase of the sintering activity of the powder. The underlying mechanisms have not been completely elucidated, but cleaning of the powder particle surfaces by dielectric effects e. g. the generation of non-conventional plasma or electromigration seems to be possible.

4. 1. 2. 6 Vacuum sintering furnace

The vacuum furnace is that the items to be heated in a vacuum environment protection sintering furnace, heating methods, such as resistance heating, induction heating, microwave heating. Figure 4. 11 shows a vacuum sintering furnace used for reaction sintering of silicon carbide ceramics in large industrial production.

Figure 4. 11 Photo of an electrically heated reaction sintering furnace

4. 1. 2. 7 Shuttle kiln

A shuttle Kiln is a car-bottom kiln with a door on one or both ends. Burners are positioned top and bottom on each side, creating a turbulent circular air flow. This type of kiln is generally a multi-car design and is used for processing whitewares, technical ceramics, and refractories in batches. Depending upon size and weight of ware, shuttle kilns may be equipped with car moving devices to transfer fired and unfired cars in and out of the kiln. Shuttle kilns can be either updraft or downdraft in design. A shuttle kiln derives its name from the fact that kiln cars can enter a shuttle kiln from either end of the kiln for processing, whereas a tunnel kiln has flow in only one direction. Figure 4. 12 shows a photo of high temperature gas-fired shuttle kiln for production of alumina ceramics.

4. 1. 3 Continuous kiln

4. 1. 3. 1 Tunnel kiln

A tunnel kiln is long tunnel-shaped furnace through which the charge is generally moved on cars, passing progressively through zones in which the temperature is maintained for

Figure 4.12　Photo of gas-fired shuttle kiln

preheating, firing, and cooling. A number of different methods can be used to heat tunnel kilns. Typical heat sources include electrical heating elements and gas-fired units, with dual heating systems also existing. Some may also conserve energy by capturing heat at the cooling end and recirculating it to help pre-heat materials that are entering the device. Figure 4.13 shows an example of a high temperature tunnel kiln firing alumina ceramic balls.

Figure 4.13　Photo of gas-fired tunnel kiln

4.1.3.2　Roller hearth kiln

Roller hearth kiln is a continuous roller conveyor type kiln that transports the products on rollers inside the kiln. It is particularly suited to the technical ceramics industry. Roller hearth kiln provides outstanding temperature uniformity, cleanliness, heat efficiency and fast firing capability. A roller hearth is available with electric heating or natural gas burners. Here we give a sample of roller hearth kiln for firing colored glaze cup, as shown in Figure 4.14.

Figure 4.14　Photo of gas-fired roller hearth kiln

4.1.3.3　Pushed slab kiln

Pushed slab kiln relates to tunnel kiln structures of the type in which conveyor slabs or plates are placed end to end and pushed through the kiln, such slabs supporting the articles or product to be fired in the kiln. A pushing slab kiln consists of kiln body, pushing slab, electrical heating components, revolving track and temperature/power control device. Figure 4.15 shows a photo of electric heating pushed slab kiln.

Figure 4.15　Photo of electric-heated pushed slab kiln

4.1.4　Heat Elements

Never use a furnace like a black box. The heating elements will go to certain temperatures, but even then you may not want that type of material close to your sample to minimize contamination. Table 4.2 lists the materials used for electrical resistance heating.

(1) Furnaces operating in air at temperatures up to 1300℃ usually use wire-wound Cr alloys.

(2) For higher temperatures in air either precious metals or SiC rods are used.

(3) For very high temperatures requiring an oxidizing environment ceramic elements, most commonly ZrO_2, are used.

(4) In cases in which reducing atmospheres can be tolerated, graphite or refractory metals such as Mo and W can be used.

We will now look in a little more detail at some of the ceramic materials that are used as heating elements in furnaces.

Table 4.2　Electrical resistance heating element materials[1]

Material	Maximum useful temperature/℃	Usable atmosphere
Chromium alloys		
Chromel C, Nichrome, Kanthal DT	1100	ONR
Kanthal A, Chromel A	1300	ONR
Metals		
Pt	1400	ONR
Pt-Rh alloys	1500-1700	ONR
Mo	1800	NR
W	2800	NR
Ceramics		
SiC	1500	ON
MoS_2	1700	ON
Lanthanum Chromite	1800	O
Thoria, stabilized	2000	ONR, shock
Zirconia, stabilizer	2800	ONR, shock
Graphite	3000	NR

Note: O, oxidizing; N, neutral; R, reducing. Shock, particularly poor resistance to thermal shock.

4.1.4.1　Silicon carbide

Silicon carbide is the most widely used non-oxide ceramic for heating elements for high-temperature furnaces. SiC heating elements can be used up to 1500℃ in air because of the formation of a protective oxide layer. At temperatures in the range 1500 ~ 1600℃ SiC decomposes:

$$SiC(s) + O_2(g) \Longleftrightarrow SiO(g) + CO(g)$$

There are three main methods used to produce SiC heating elements: in situ reaction, reaction bonding and sintering.

In the first method a carbon tube is heated to about 1900℃ in a bed of sand (SiO_2) and coke (C). The tube may be directly resistance heated or heated indirectly by a sacrificial tube of smaller diameter. Silicon monoxide is generated and infiltrates the carbon tube transforming it to SiC. The SiC tube is then removed and the residual carbon is burnt out. The tube has a porosity of about 30% and a large internal surface area. To prevent internal oxidation during use the outer surfaces of the tube are coated with a thin layer of a calcium aluminosilicate glass and are then fired at about 1450℃. In this form the tubes have a uniform resistance along their length. A higher-resistance heating section is made by diamond sawing a spiral through the tube wall as shown in Figure 4.16. Adjusting the pitch of the cut varies the resistance.

Figure 4.16 Examples of SiC furnace elements

In the second method a mix of SiC and carbon powders and a polymer binder is extruded to a rod. The "green" form is then brought in contact with molten silicon. The liquid penetrates the pores reacting with the carbon to form silicon carbide and bonding the grains together. The resulting ceramic has low porosity and, consequently, a long service life. The resistance of the hot section of the rod is adjusted to the required value by spiraling, which is easier to do when the ceramic is in the "green" state.

In the third method, SiC powder is mixed with a polymer binder and extruded. The rod is then sintered in a carbon furnace at approximately 2300℃. To give the rod low-resistance terminations the ends are dipped into molten silicon, which is allowed to infiltrate along a pre-determined length. In all cases the ends of the elements are coated with aluminum to make electrical contacts.

Silicon carbide elements are available in a wide range of shapes and sizes. However, the main disadvantage of silicon carbide heating elements is that they are extremely brittle and must be handled carefully, especially when being installed and wired.

4.1.4.2　Molybdenum disilicide

Many metals form conductive silicides that, like SiC, are resistant to oxidation through the formation of stable layers of silicates or silica on their surfaces at high temperatures. Molybdenum disilicide ($MoSi_2$) has been developed as a heating element for use in air at temperatures above 1500℃. The resistivity of $MoSi_2$ behaves in the same way as for a metal—it increases with increasing temperature. The room-temperature resistivity of $MoSi_2$ is $2.5 \times 10^{-7} \Omega \cdot m$; it increases to about $4 \times 10^{-6} \Omega \cdot m$ at 1800℃.

A commercial $MoSi_2$ heating element, known as Kanthal Super, is composed of a mixture of $MoSi_2$ particles bonded together with an aluminosilicate glass phase, which forms 20% of the total volume. The elements are fabricated by extruding a mixture of fine $MoSi_2$ powder with clay. The rods are dried, sintered, and cut to various lengths. The heating zones are bent to the required shape at high temperature and are then welded to the larger-diameter terminal sections. The best grade of $MoSi_2$ element is capable of operating at temperatures up to 1800℃.

4.1.4.3　Graphite

Graphite is a good choice as a heating element for resistive heating because it has a high melting temperature and a very low vapor pressure even at temperatures above 3000℃. These characteristics led to the use of graphite filaments in early incandescent lamps. (They were eventually replaced by tungsten, which came into general use as a filament for incandescent lamps in 1911). Reactivity with oxide ceramics and susceptibility to oxidation are the major disadvantages of using graphite furnace elements.

All metal oxides are reduced when in direct contact with graphite at high temperatures. Even the most refractory oxides will be reduced if the temperature is high enough.

4.1.5　Insulation and Refractories

A thermal insulator is a poor conductor of heat and has a low thermal conductivity. Insulation is used in manufacturing processes to prevent heat loss or heat gain. Such materials are porous, containing large number of dormant air cells.

Refractories are materials capable of withstanding high temperatures and not degrading in a furnace environment when in contact with corrosive liquids and gases. Refractory insulators are used in high-temperature applications to reduce heat losses and to save fuel. Table 4.3 lists some of the important furnace insulation materials together with their maximum usable temperature and thermal conductivity, κ. The lower the value of κ, the better the thermal insulating effect for equal thickness.

Table 4.3   Refractories for thermal insulators[1]

Material	T_{max}/℃	κ/W · (m · K)$^{-1}$
SiO_2, fiber	1000	0.17
Firebrick, insulating	1200~1500	0.52
Al_2O_3, bubble	1800	1.04
MgO, powder	2200	0.52
Carbon or graphite, powder	3000	0.09

We have invented two kinds of composite thermal insulators based on potassium hexatitanate whisker with aluminium silicate fiber and basalt fiber respectively[8,9]. For the potassium hexatitanate whisker/aluminium silicate fiber thermal insulator, the thermal conductivity at 800℃ is 0.77W/(m · K), it can be used below 1200℃. The potassium hexatitanate whisker/basalt fiber thermal insulator can be used below 1000℃, with a thermal conductivity at 800℃ 0.052W/(m · K).

4.1.6 Furniture and Crucibles

Crucibles and other furnace equipment such as boats and setter plates must meet the same requirements as refractory materials used for furnace insulation, i.e., they must be able to withstand high temperatures and also contact with any corrosive liquids or gases used. Items such as crucibles and boats should also possess good thermal shock resistance as they may be heated and cooled rapidly. Figure 4.17 shows an example of kiln furniture used for the production of Al_2O_3 ceramic ball. These plate and stand column are made from Si_3N_4 bonded SiC. Any component in contact with a crucible or other piece of furnace equipment at high temperature can be contaminated.

Recently, in order to save energy, a novel kiln furniture and set method are being used in the production of dinnerware, as shown in Figure 4.18. These SiC bars replaced original cordierite plates. With this kind of furniture, the kiln can save energy over 40%.

Figure 4.17　Kiln furniture based on Si_3N_4 bonded SiC

Figure 4.18　A novel kiln furniture

4.1.7　Measuring Temperature

In this section we will describe some of the approaches used to determine the high temperatures employed in the processing of ceramics.

4.1.7.1　Thermocouples

The most common and convenient means of measuring temperature is to use a thermocouple. In 1821 a German physicist named Seebeck discovered the thermoelectric effect which forms the basis of modern thermocouple technology. He observed that an electric current flows in a closed circuit of two dissimilar metals if their two junctions are at different temperatures. The thermoelectric voltage produced depends on the metals used and on the temperature relationship between the junctions. If the same temperature exists at the two junctions, the voltages produced at each junction cancel each other out and no current flows in the circuit. With different temperatures at each junction, different voltages are produced and current flows in the circuit. A thermocouple can therefore measure temperature differences between the two junctions. In Seebeck's original research the two metals were bismuth and copper.

There are many different types of thermocouples available to cover temperatures ranging from −273℃ to 2000℃. The most important ones are given in Table 4.4. The type of thermocouple you would use depends mainly on the temperature range over which you will be requiring information and the desired accuracy of the temperature reading. The most commonly used thermocouples for ceramic applications are types K, R, and C. These are used for temperatures up to 1250℃, 1450℃, and 2300℃, respectively. At lower temperature it is preferable to use a base metal combination such as chromel-alumel, which gives greater accuracy.

Table 4.4 Characteristics of thermocouples[1]

Type	Combination of metals or alloys	Output① at 900℃/mV	Temperature limit/℃	Applications
T	Cpper-constantan	20.9②	400	Mild oxidizing, reducing, vacuum, or inert. Good where moisture is present. Low temperature and cryogenic applications
J	Iron-constantan	21.9②	760	Reducing, vacuum, inert. Limited use in oxidizing at high temperature. Not recommended for low temperature
E	Chromel-constantan	68.8	900	Oxidizing or inert. Limited use in vacuum or reducing. Highest EMF change per degree
K	Chromel-alumel	37.3	1250	Clean, oxidizing or inert. Limited use in vacuum or reducing. Wide temperature range. Most stable at high temperature
S	Pt-Pt 10% Rh	8.4	1450	Alternative to Type K. More stable at high temperature. Oxidizing or inert. Do not insert into metal tubes. Beware of contamination. High temperature
R	Pt-Pt 13% Rh	9.2	1450	Same as Type S
B	Pt6% Rh-Pt 30% Rh	4.0	1700	Oxidizing or inert. Do not insert into metal tubes. Beware of contamination. High temperature. Common use in glass industry

Continues Table 4.4

Type	Combination of metals or alloys	Output[1] at 900℃/mV	Temperature limit/℃	Applications
G(W)	W-W 26% Re	12.3	2300	Vacuum, inert, hydrogen. Beware of embrittlement. Not practical below 750℃. Not for oxidizing atmosphere
C(W5)	W 5% Re-W26% Re	16.4	2300	Same as Type G. Nonoxidizing atmosphere only
D(W3)	W 3% Re-W 25% Re	15.1	2300	Same as Type G. Nonoxidizing atmosphere only

[1] The higher the output voltage, the simpler the associated circuitry and/or the more accurate the temperature reading. [2] Output at 400℃.

In furnaces, the leads of the thermocouple are usually isolated from each other and other parts of the furnace by placing them either in thin alumina sheaths or by threading alumina beads along their length. The thermocouple should ideally be placed directly into the furnace cavity close to the object being heated. The external circuitry that measures temperature and, through an associated electrical circuit, controls the power to the heating elements is kept outside the furnace.

A thermocouple is a very accurate means of measuring temperature. But you should always bear in mind that the temperature being measured is that at the thermocouple tip. Unless the thermocouple is in intimate contact with the ceramic parts being heated, they may actually be at a different temperature than that measured by the thermocouple. A good illustration of this point is that thermocouples are often used to give substrate temperatures during thin film growth. The thermocouple is frequently attached to be substrate support but is not ill direct contact with the substrate itself. In controlled tests it has been found that the measured thermocouple temperature and the actual surface temperature of the substrate can be off by as much as 100℃ or in some cases even more.

4.1.7.2 Pyrometers

At any temperature, above 0 K, all objects emit electromagnetic radiation in accordance with the Stefan-Boltzmann law. It is possible to estimate the temperature of a hot object by its color. In 1557 in his work on Renaissance pottery, Cipriano Piccolpasso described how the furnace operator was able to use color variations to judge furnace temperatures. As the temperature of an object increases the range of wavelengths that it emits increases and shifts to shorter values. At temperatures above 500℃ there will be a red coloration that will become increasingly orange as the temperature increases beyond 1000℃. At 1400℃ the object will appear bright white. Be careful when looking at hot objects and remember that an object can still be too hot to handle even if it is not glowing red[1].

Optical pyrometers allow direct measurement of the temperature of an object. The disappearing filament optical pyrometer works by matching the intensity of radiant energy coming from an incandescent source to the intensity of a calibrated filament when both the source and the filament are viewed through a red filter. When the filament and the source intensities are the same, the image of the filament disappears as it is superimposed on the image of the source. An obvious requirement of this technique is that it can be used only to measure temperatures that produce visible incandescence, above 750℃.

The advantages of the disappearing filament pyrometer are as follows:

(1) The distance from the target to the source is not important (it is only necessary to have a clear image to compare with the filament image).

(2) Various places on a target can be examined for temperature distribution.

(3) Very high temperatures can be measured.

(4) The visible image ensures that the instrument is measuring the temperature of the desired portion of the target.

Reasonable accuracy (at best ±0.2℃ at 775℃ reduced to ±1℃ at 1225℃) can be obtained.

The disadvantages are as follows:

(1) The instrument either must be sighted under blackbody conditions or the reading corrected for emittance.

(2) Absorption by dust, windows, flame, and other optical interference can produce errors.

(3) The disappearing filament optical pyrometer is slow and manual. However, other pyrometers such as the photoelectric optical pyrometer can be automated.

4.1.7.3 Pyrometric cones

Pyrometric cones are pyrometric devices that are used to gauge heatwork during the firing of ceramic materials. The cones, often used in sets of three as shown in Figure 4.19, are positioned in a kiln with the wares to be fired and provide a visual indication of when the wares have reached a required state of maturity, a combination of time and temperature. Thus, pyrometric cones give a temperature equivalent, they are not simple temperature-measuring devices.

The pyrometric cone is "a pyramid with a triangular base and of a defined shape and size"; the "cone" is shaped from a carefully proportioned and uniformly mixed batch of ceramic materials so that when it is heated under stated conditions, it will bend due to softening, the tip of the cone becoming level with the base at a definitive temperature. Pyrometric cones are made in series, the temperature interval between the successive cones usually being 20 ℃. The best known series are Seger Cones (Germany), Orton Cones (USA) and Staffordshire Cones (UK).

Figure 4.19 Orton cones: the self-supporting type
(The cones shown are 6 cm tall in their initial state)

Since pyrometric cones are sensitive to both time and temperature, the actual temperature associated with each cone can vary, but this is also one of the reasons why they are very useful for ceramic processing. Sintering, for example, depends on both time and temperature.

Pyrometric rings are flat, hollow centred rings whose contraction is proportional to the heat work experienced. A micrometer or gauge measures the fired ring, with the difference being an arbitrary number that is used to describe the firing regime experienced. Various grades of ring, each of slightly different compositions, are available to cover all firing conditions and temperature equivalents likely to be encountered. Examples of pyrometric rings include Bullers Rings and Philips Rings. Pyrometric rings are the best way of measuring temperature/time in all ceramic firing methods. No clay body support is required and a single ring is suitable for carrying out several different temperature measurements. Furthermore, it is not necessary to remove the firing material in order to position them. Temperature control is therefore made much easier, as shown in Figure 4.20.

Figure 4.20 Bullers firing trial rings

4.2 Sintering of Advanced Ceramics

The sintering process converts the green microstructure to the microstructure of the dense ceramic component. In this way, sintering is the last of the ceramic processing steps where the ceramist has an influence on microstructural development. This influence is limited, however, as the worst in homogeneities that pre-exist in the compact are usually exaggerated during sintering; for example, flaws will persist or even grow, while large particles may induce abnormal grain growth[10].

The sintering process consists of solid particle bonding or neck formation, followed by continuous closing of pores from a large open porosity to essentially porefree bodies. There are various sintering processes which occur by different mechanisms. Traditional household and sanitary ceramic ware are densified by viscous flow. In contrast, advanced ceramics are produced by **liquid-phase** and **solid-phase sintering**, which utilize significantly smaller amounts of sintering additives as compared to viscous flow densification. Liquid-phase sintering involves solution-reprecipitation and diffusion mechanisms, while solid-state sintering is dominated by volume and grain boundary diffusion mechanisms responding to free energy and chemical potential differences. Solid state sintering occurs when the powder compact is densified wholly in a solid state at the sintering temperature, while liquid phase sintering occurs when a liquid phase is present in the powder compact during sintering. Viscous flow sintering occurs when the volume fraction of liquid is sufficiently high, so that the full densification of the compact can be achieved by a viscous flow of grain-liquid mixture without having any grain shape change during densification. Transient liquid phase sintering is a combination of liquid phase sintering and solid state sintering. Figure 4.21 illustrates the various cases in a

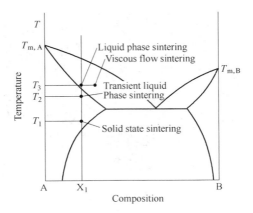

Figure 4.21　Illustration of various types of sintering[3]

schematic phase diagram. At temperature T_1, solid state sintering occurs in an A-B powder compact with composition X_1, while at temperature T_3, liquid phase sintering occurs in the same powder compact. Transient liquid phase sintering may also be found when the A-B powder compact with composition X_1 is sintered above the eutectic temperature but below a solidus line, at temperature T_2[3].

Another method of producing advanced ceramics is the **reactive sintering processes**. Here, the solid phase precipitates from an intermediary liquid phase which is generated during compaction.

When the particle size of the precursor powders decreases to nanosized powders, new mechanisms such as grain boundary slip, dislocation motion, grain rotation, viscous flow and grain boundary melting become operative. Furthermore, sintering of nanopowders enjoys a high driving force and enhanced kinetics due to the curvature effect. Thus, the densification of nanopowders occurs at temperatures significantly below those of larger-grained powders by up to several hundreds of degrees. Consequently, small final grain sizes may result and sintering aids and undesirable phase transformation may be avoided.

For the enhanced densification of ceramic powders, pressure-assisted consolidation methods such as **hot pressing**, **hot isostatic pressing**, **sinter forging**, **hot extrusion** or **ultra-high pressure sintering** can be applied. Hot pressing is a technique which combines external uniaxial pressure with temperature in order to enhance densification. The hot isostatic pressing (HIP) technique combines high temperature and a **gas pressure**, which is uniformly applied to the powders in all directions.

Besides these methods, a number of non-conventional consolidation techniques have been applied to ceramic powder sintering, including **microwave sintering** and **spark plasma sintering**.

In this section, we will introduce pressureless sintering, hot press sintering, gas pressure sintering, reaction sintering, microwave sintering and spark plasma sintering. First of all, we introduce solid-sintering and liquid-sintering briefly, because whether in hot pressing sintering or in microwave sintering or in spark plasma sintering, it is possible where exists solid-sintering or liquid sintering which depends on the amount of liquid during sintering.

4.2.1 Solid-state Sintering and Liquid-sintering

4.2.1.1 Solid-state sintering

Solid state sintering is usually divided into three overlapping stages—initial, intermediate and final. Figure 4.22 schematically depicts the typical densification curve of a compact through these stages over sintering time. The initial stage is characterized by the formation of necks between particles and its contribution to compact shrinkage is limited to 2%~3% at most. During the intermediate stage, considerable densification, up to about 93% of the relative density, occurs before isolation of the pores. The final stage involves densification from the isolated pore state to the final densification.

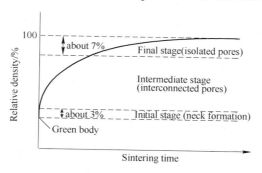

Figure 4.22 Schematic showing the densification curve of a powder compact and the three sintering stages[3]

For each of these three stages, simplified models are typically used: the two-particle model for the initial stage, the channel pore model for the intermediate stage, and the isolated pore model for the final stage. Although all models ignore grain growth during sintering, they do provide a means of analysing the densification process and evaluating the effects of various processing parameters.

Figure 4.23 shows two geometrical models for two spherical particles: one without shrinkage (a) and the other with shrinkage (b). In Figure 4.23a, the distance between the particles does not change but the neck size increases as the sintering time increases. In the model with shrinkage (Figure 4.23b), the neck size increases with an increased sintering time by material transport between the particles and hence shrinkage results. When necks form between particles in real powder compacts, pores form interconnected channels along 3-grain edges. As the sintering proceeds, the pore channels are disconnected and isolated pores form because the dihedral angle is much larger than 60° and the shrinkage of the interconnected pores is not uniform due to the non-constant size of the pore channels and also Rayleigh's surface instability. It is assumed to last until the radius of the neck between the particles has reached a value of about 0.4 ~ 0.5 of the particle radius. For a powder system with an initial density of 0.5 ~ 0.6 of the T.D., this corresponds to a linear shrinkage of 3% to 5% or an increase in density to about 0.65 of the theoretical when the densifying mechanisms dominate.

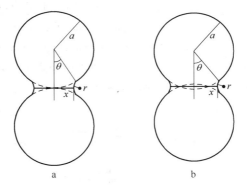

Figure 4.23 Two-particle model for initial stage sintering without shrinkage (a) and with shrinkage (b)[3]

Coble's microstructure model of intermediate stage sintering is based on bcc-packed tetrakaidecahedral grains with cylinder-shaped pores along all of the grain edges, as shown in Figure 4.24a. This intermediate stage model assumes equal shrinkage of pores in a radial direction. Densification is assumed to occur by the pores simply shrinking to reduce their cross section. Eventually, the pores become unstable and pinch off, leaving isolated pores; this constitutes the beginning of the final stage. Although the model is limited in terms of describing real sintering, it reasonably simplifies sintering complexity and allows the evaluation of the effect of sintering variables on sintering kinetics.

The microstructure in the final stage can develop in a variety of ways. In one of the

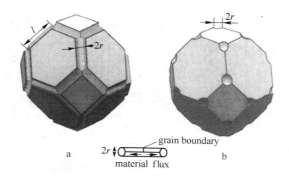

Figure 4.24 Coble's geometricalmodels for intermediate stage (a) and final stage sintering (b)[3]

simplest descriptions, the final stage begins when the pores pinch off and become isolated at the grain corners, as shown by the idealized structure in Figure 4.24b. In this simple description, the pores are assumed to shrink continuously and may disappear altogether. The removal of almost all of the porosity has been achieved in the sintering of several real powder systems[3].

Some of the main parameters associated with the three idealized stages of sintering are summarized in Table 4.5, and examples of the microstructures (planar section) of real powder compacts in the initial, intermediate, and final stages are shown in Figure 4.25.

Table 4.5 Parameters associated with the stages of sintering for polycrystalline solids[11]

Stage	Typical microstructural feature	Relative density range	Idealized model
Initial	Rapid inter-particle neck growth	Up to 0.65	Two monosize spheres in contact
Intermediate	Equilibrium pore shape with continuou porosity	0.65 ~ 0.90	Tetrakaidecahedron with cylindrical pores of the same radius along the edges
Final	Equilibrium pore shape with isolated porosity	>0.90	Tetrakaidecahedron with spherical monosize, pores at the corners

Many ceramics have been densified by solid state sintering, especially the relatively pure oxides. Examples include BeO, Y_2O_3, UO_2, ThO_2, ZrO_2 and doped ZrO_2. Doping can increase the number of point defects in the material and increases the rate of diffusion, thus enhancing solid-state sintering. SiC with the addition of B and C is also

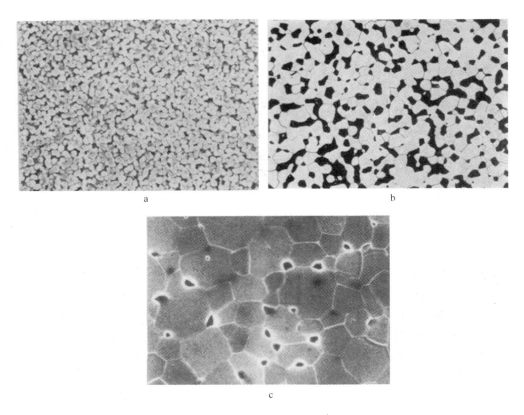

Figure 4.25 Examples of real microstructures (planar sections) for initial stage of sintering (a), intermediate stage (b), and final stage (c)[11]

thought to densify by solid-state sintering. The C apparently removes SiO_2, from the surface of the SiC particles. The B has limited solid solubility in the SiC and allows a mechanism of material transfer between adjacent grains. Pure SiC particles can bond together, but not densify (no shrinkage and no removal of interparticle porosity).

4.2.1.2 Liquid-phase sintering

The purpose of liquid-phase sintering is to enhance densification rates, achieve accelerated grain growth, or produce specific grain boundary properties. The distribution of the liquid phase and of the resulting solidified phases produced on cooling after densification is critical to achieving the required properties of the sintered material. Commonly, the amount of liquid formed during sintering is small, typically less than a few volume percent (%), which can make precise control of the liquid composition difficult. In some systems, such as Al_2O_3, the amount of liquid phase can be very small and so difficult to detect that many studies that were believed to involve solid-state sintering actually involved liquid silicate phases, as later revealed by careful high-resolution transmission electron microscopy[11].

Liquid-phase sintering is particularly effective for ceramics such as Si_3N_4 and SiC that have a high degree of covalent bonding and are therefore difficult to densify by solid-state sintering. The process is also important when the use of solid-state sintering is too expensive or requires too high a fabrication temperature. However, the enhanced densification rates achieved by liquid-forming additives are only of interest if the properties of the fabricated ceramic remain within the required limits. A disadvantage of liquid-phase sintering is that the liquid phase used to promote sintering commonly remains as a glassy intergranular phase that may degrade high temperature mechanical properties such as creep and fatigue resistance.

Some examples of ceramic liquid-phase sintering systems and their applications are given in Table 4.6.

Table 4.6 Examples of common ceramic liquid-sintering systems[11]

Ceramic systems	Additive content (mass fraction)/%	Application
Al_2O_3 (talc)	about 5	Electrical insulators
$ZnO(Bi_2O_3)$	2~3	Electrical varistors
$BaTiO_3(TiO_2)$	<1	Dielectrics
$BaTiO_3(LiF)$	<3	Dielectrics
$UO_2(TiO_2)$	about 1	Nuclear ceramics
$ZrO_2(CaO\text{-}SiO_2)$	<1	Ionic conductors
$Si_3N_4(MgO)$	5~10	Structural ceramics
$Si_3N_4(Y_2O_3\text{-}Al_2O_3)$	5~10	Structural ceramics
$SiC(Y_2O_3\text{-}Al_2O_3)$	5~10	Structural ceramics

A related process is activated sintering in which minor amounts of additives that segregate strongly to the grain boundaries can significantly enhance mass transport rates along the grain boundary, giving rise to accelerated densification even at temperatures well below that for liquid formation in the system. In many systems, there is no clear difference in principles between activated sintering and liquid-phase sintering, except that for the activated system, the amount of additive is fairly small so that the presence of a liquid grain boundary film can be difficult to detect. If sufficient liquid is present (on the order of 25% ~ 30% (mass fraction)), rearrangement of the solid phase coupled with liquid flow can lead to a fully dense material. Such large volume fractions of liquid are commonly used in traditional, clay based ceramics such as porcelains and in cemented carbides. In the traditional ceramics, the liquid phases are molten silicates that remain as a glassy phase after cooling, giving the fabricated materials a glassy appearance. The ceramics are referred to as vitrified, and the sintering process is referred to as vitrification.

In most liquid-phase sintering systems, chemical reactions between the particulate solid and the liquid are relatively weak, so that the interfacial energies have a dominant effect on the rate of sintering. Under these conditions, as illustrated in Figure 4.26, liquid-phase sintering is generally regarded as proceeding in a sequence of dominant stages:

(1) Redistribution of the liquid and rearrangement of the particulate solid under the influence of capillary stress gradients.

(2) Densification and grain shape accommodation by solution-precipitation.

(3) Final-stage sintering driven by the residual porosity in the liquid.

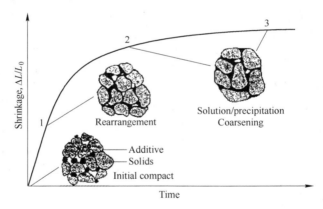

Figure 4.26 Schematic evolution of a powder compact during liquid-phase sintering
(The three dominant stages overlap significantly[11])

As the temperature of the green body is raised, solid-state sintering may occur prior to the formation of the liquid, producing significant densification in some systems. Assuming good wetting between the liquid and the particulate solid, further densification occurs as a result of the capillary force exerted by the liquid on the particles. The particles shrink as solid dissolves in the liquid and rapidly rearrange to produce a higher packing density, releasing liquid to fill pores between the particles. Capillary stresses will cause the liquid to redistribute itself between the particles and into the small pores, leading to further rearrangement. Contact points between agglomerates will be dissolved due to their higher solubility in the liquid, and the fragments will also undergo rearrangement. Throughout the process, dissolution of sharp edges will make the particle surfaces smoother, thereby reducing the interfacial area and aiding the rearrangement of the system. Initially, rearrangement occurs rapidly, but as densification occurs, the viscosity of the system increases, causing the densification rate to decrease continuously.

As densification by rearrangement slows, effects dependent on the solid solubility in the liquid and the diffusivity in the liquid dominate, giving the second stage termed solution-precipitation. The solid dissolves at the solid-liquid interfaces with a higher chemical potential, diffuses through the liquid, and precipitates on the particles at other sites with a lower chemical potential. One type of dissolution site is the wetted contact area between the particles where the capillary stress due to the liquid or an externally applied stress leads to a higher chemical potential. Precipitation occurs at sites away from the contact area. For systems with a distribution of particle sizes, matter can also be transported from the small particles to the large particles by diffusion through the liquid. The net result is a coarsening of the microstructure.

The final stage of liquid-phase sintering is controlled by the densification of the solid particulate skeletal network. The process is slow because of the large diffusion distances in the coarsened structure and the rigid skeleton of contacting solid grains. The residual pores become larger if they contain trapped gas, leading to compact swelling. Coarsening is accompanied by grain shape accommodation, which allows more efficient packing of the grains. Liquid may be released from the more efficiently packed regions and may flow into the isolated pores, leading to densification.

The stages of liquid phase sintering are summarized in Figure 4.26. The extent to which each stage influences densification is dependent on the volume fraction of liquid so there are many variants in this conceptual picture. When the volume fraction of liquid is high, complete densification can be achieved by the rearrangement process alone. On the other hand, at the low liquid contents common for many systems, the solid skeleton inhibits densification, so that solution-precipitation and final stage sintering are required to achieve further densification[11].

Si_3N_4-based compositions represent an advanced family of ceramics that are densified by liquid-phase sintering. Additives are required to achieve a liquid phase. Examples of additives that have been used successfully include MgO, Al_2O_3, Y_2O_3, and rare earth oxides. There react with the SiO_2 present on the surface on the Si_3N_4 particles to form a silicate liquid. The Si_3N_4 particles dissolve in the liquid and precipitate as β-Si_3N_4 crystals. If the starting powder is α-Si_3N_4 and if the time and temperature cycle is properly selected, the β-Si_3N_4 will crystallize as elongated single crystals that impart unusually high fracture toughness to the Si_3N_4. Figure 4.27 shows the microstructures of a Si_3N_4-Y_2O_3-Al_2O_3 composition sintered at different temperatures. 1600℃ was too low to achieve full densification. Much porosity is present and the original Si_3N_4 particles are still visible and of the α-Si_3N_4 phase, 1750℃ resulted in nearly complete densification and conversion of the Si_3N_4 to elongated β-Si_3N_4, 1850℃ resulted in substantial grain growth[2].

a　　　　　　　　　　　　　　b　　　　　　　　　　　　　　c

Figure 4.27　Si_3N_4 containing Y_2O_3 + Al_2O_3 sintering aids; sintered in nitrogen at 1600℃ (a), 1750℃ (b) and 1850℃ (c)[2]

4.2.2　Pressureless Sintering

Pressureless sintering is the sintering of a powder compact (sometimes at very high temperatures, depending on the powder) without applied pressure. Here we will emphasize the sintering of silicon carbide and boron carbide.

4.2.2.1　Pressureless sintering of silicon carbide ceramics

Silicon carbide (SiC) is a very interesting ceramic material due to its properties like high hardness, low bulk density, high oxidation resistance which made SiC useful for a wide range of industrial application. The main reasons for the attention focused on ceramics as possible structural materials are their heat resistance and wear resistance. Application of SiC compacts as structural materials can be broadly divided into two categories: (1) abrasion and corrosion resistant components; (2) heat resistant components. It was found that SiC sintered bodies possess superior strength at high temperatures, and they were seen as a promising heat-resistant structural material. Their characteristics include high temperature properties like creep resistance, bending strength, high hardness, heat resistance and corrosion resistance. These properties enable SiC to be the prime material for structural components such as high temperature furnace components (heating elements, core tubes, refractory bricks, etc). These applications, however, do not require high density sintered bodies. The developments of SiC compacts as structural components with high density were initially begun with the objective of application to gas turbine engines.

Si-C bonds have a high degree of covalency (87%) and this is the source of the intrinsically high strength of SiC sintered bodies. SiC is fundamentally a brittle material. Its low fracture toughness presents difficulties in application as a structural material. But SiC sintered bodies belong to the few materials, which can be used at extremely high temperatures (e.g. 1300℃ and above). However, due to the highly covalent bond characteristics, it is difficult to sinter these materials to full density. The poor sinterability necessitates addition of a sintering aid for densification. Addition of suitable sintering additives leads to dense, fine-grained microstructures and, hence, high strength of the sintered body, but generally it may also result in the formation of secondary phases at the grain boundaries that frequently cause loss of high temperature strength due to softening of these phases. Three representative technologies have been developed for sintering of SiC namely, **hot pressing** with a sintering aid, **pressureless** or **gas pressure sintering** with a sintering aid, and **reaction sintering**. Among these techniques, pressureless sintering is considered to be the most important from an industrial standpoint because it allows the manufacture of large or complex-shaped pieces, offers good mass productivity and low cost, and can produce products with superior performance[12].

Sintering of silicon carbide was first performed by Prochazka[13] by using boron and carbon as sintering aids. It is opined that very fine particle size along with B and C additives would promote the sinterability by enhancing volume diffusion and retarding surface diffusion. The mechanisms was removal of the surface oxide layer of silicon carbide by carbon and reduction of the grain boundary energy by segregation of boron. Additions of boron and carbon play an important role not only in densification but also in the inhibition of the grain growth of SiC crystals. Carbon and boron inhibit silicon carbide grain growth caused by surface diffusion during heating up to the sintering temperature, thereby maintaining the driving force for sintering up to higher temperatures where volume diffusion-induced densification becomes active. Addition of carbon or boron alone is not effective for such an activated sintering mechanism. D. H. Stutz et al[14] Studied the effect of boron with carbon and aluminium with carbon in the sintering of β-SiC and found that a minimum B concentration of 0.3% (mass fraction) was required to promote densification. Carbon must be present to the tune of 2% (mass fraction). The effectiveness of aluminium as a SiC sintering additive was discovered by R. A. Alliegro et al[15] who observed that aluminium was the most effective additive after boron. But addition of only 1% Al with 1.5% C could not densify the cubic SiC powders to high density.

Giuseppe Magnani et al[16] studied the pressureless sintering process of SiC-based material containing boron carbide as secondary phase (5% (volume fraction)). Boron carbide addition permitted to increase sintering rate of SiC powder to reach a 96% T. D. at 2150℃, reduced grain growth, and increased the flexural strength up to 1500℃. These properties make this composite a very interesting material for application in which high oxidation resistance, e. g. ceramic exchanger.

Apart from the additives, the **oxygen content** of sinterable silicon carbide powders also affects their densification behaviour[17]. Lower oxygen content improves the densification behaviour of SiC. It was observed that 9% difference in sintered density existed between a standard powder and a heat treated powder with 0.64% (mass fraction) O_2. The morphology of the oxygen containing phases had an influence on the densification. Powders having higher specific surface area had a higher densification rate. Silicon carbide sintered bodies with a relative density of 95% or more were produced by pressureless sintering at 2050℃ if an ultrafine raw powder with particle size of less than 0.05μm was used[18]. Control of **sintering atmosphere** and heating rate were also important for the sintering of silicon carbide powders. Inert gas atmosphere gave optimum results with B-doped powders. Higher pressure of the inert gas atmosphere reduced the sintered density. N_2 had a retarding effect on the sintering of boron doped β-SiC. Silicon vapour, oxygen and CO_2 gas in the system were deleterious to sintering as they resulted in a loss of boron and consequently in the inhibition of densification. Sintering was strongly inhibited by silicon vapour as a result of coarsening of the compacts. A rapid heating rate was not favourable and in the presence of some of the additives, selection of a slower heating rate was the most effective way to increase the sintered density of silicon carbide.

Besides solid state sintering, one of the pressureless sintering methods is liquid phase sintering (LPS) which has been applied extensively to obtain highly dense (>95% T. D.) SiC ceramics. Several additives or combinations of them, such as Al_2O_3-Y_2O_3 (or rare-earth oxides), Al_2O_3-Y_2O_3-CaO, YAG and AlN-Y_2O_3 (or rare earth oxides) have been investigated for SiC in order to improve the thermomechanical behaviour.

In LPS-SiC materials, typically 5% ~ 15% (mass fraction) of the Al_2O_3-Y_2O_3 additive system is commonly used. It was reported[19] that the sintering rate increased with increasing amount of sintering additives in the Al_2O_3-Y_2O_3 system. The sintering temperatures used for the Al_2O_3-Y_2O_3 system are lower in Ar than in N_2 atmosphere. This results in less grain growth in Ar-sintered samples and there is also more mass-loss when sintering in N_2, which makes Ar a more favourable gas atmosphere to sinter.

Giuseppe Magnani et al[20] studied the properties of liquid phase pressureless sintering SiC obtained without sintering bed. They successfully liquid pressureless sintered SiC using 4% (mass fraction) Y_2O_3 and 6% (mass fraction) Al_2O_3 as sintering aids in the temperature range 1850 ~ 1950℃, dwell time of 0.5 h, in flowing Ar at 1 atm. The sintered density reached the value of 97% T. D. at 1875℃. Furthermore, this material exhibits high fracture toughness and high oxidation resistance up to 1400℃ that makes it potentially promising for a wide range of applications also at high temperature.

E. Gomezl et al[21] reported the liquid phase sintering of SiC with additions of Y_2O_3, Al_2O_3 and SiO_2. Their research results indicated that pressureless sintering of SiC based ceramics in the presence of a liquid phase leads to fully dense materials provided that the liquid formed at the sintering temperature has adequate characteristics, as those formed by sintering aids like alumina and yttria. Additionally, the liquids formed by these sintering aids solidify as crystalline intergranular phases forming the yttrium aluminium garnet (YAG). The adding SiO_2 to the powder mixture has a negative effect on the densification behaviour of silicon carbide based ceramics. Carbon was revealed as an important sintering aid for this type of ceramics, promoting sintering by reducing the viscosity of the liquid and through the reduction of SiO_2, thus improving wetting. In contrast, carbon encourages surface transport mechanisms, thus causing grain growth of the 4H hexagonal polytype at the highest temperatures.

T. Nagano and K. Kaneko[22] reported liquid phase sintering SiC using AlN together with yttria and/or alumina. During liquid phase sintering, AlN forms low-melting eutectic composition with Y_2O_3 and/or Al_2O_3. When N_2 is used as sintering atmosphere, silicon carbide sintered with the Y_2O_3-AlN system has the advantage that decomposition reaction during sintering can be suppressed by application of a nitrogen overpressure. Moreover, N_2 atmosphere was not only able to improve the densification by suppressing the decomposition reaction of AlN, but also beneficial for producing fine-grained SiC with high strength since the high viscosity of N-containing melts retard the $\beta \rightarrow \alpha$-SiC phase transformation and inhibit grain growth. The glass transition temperatures and softening points of these N-enriched secondary phases (mostly oxynitrides) are significantly higher than those of silicate glasses, contributing to superior high-temperature properties of the sintered ceramics.

Xingzhong Guo et al[23] studied the sintering behavior, microstructure and mechanical properties of silicon carbide ceramics containing different nano-TiN additive. The silicon carbide ceramics (with 10% (mass fraction) Al_2O_3 and Y_2O_3 mixture at a molar ratio of Y : Al = 3 : 5) with nano-TiN additive were liquid phase sintered at 1950℃ for 15 min and subsequently sintered at 1850℃ for 1 h. The results show that the addition of nano-TiN particles restrains the densification of silicon carbide ceramic, inhibits the grain growth of ceramic. And the reactions of TiN with SiC and Al_2O_3 to form new phases of TiC and AlN may benefit the silicon carbide ceramics in a certain range of nano-TiN addition. The silicon carbide ceramic with 5% (mass fraction) of nano-TiN possesses high densification, uniform microstructure and superior mechanical properties. The toughening mechanism of nano-TiN on silicon carbide ceramic mainly comes from thermal residual stresses, crack deflection and crack bridging.

F. Rodríguez-Rojas et al[24] studied the oxidation behaviour of pressureless liquid-phase-sintered SiC with additions of $5Al_2O_3 + 3RE_2O_3$ (RE = La, Nd, Y, Er, Tm or Yb). It was found that the oxidation is in all cases passive and protective, with kinetics governed by the arctan-rate law. This is because the PLPS SiC ceramics develop oxide scales having no cracks or open porosity and accordingly prevent the parent material from direct contact with oxygen. In addition, these oxide scales crystallize gradually during the exposure to the oxidizing atmosphere with the attendant reduction in the amorphous cross-section available for oxygen diffusion. It was also found that the rate-limiting mechanism of the oxidation is outward diffusion of RE^{3+} cations from the intergranular phase into the oxide scale, and that the activation energy of the oxidation increases with increasing size of the RE^{3+} cation. It was also observed that the oxidation of PLPS SiC increases with increasing size of the RE^{3+} cation, an effect that is especially marked for cation sizes above 0.09nm because the oxidation rate becomes several orders of magnitude faster. This trend is attributable to the oxide scales being more crystalline, and containing crystals that are more refractory and amorphous residual phases that are more viscous as the size of the RE^{3+} cation decreases.

Typically LPS-SiC ceramics consist of highly refractory SiC grains bonded by a less refractory vitreous phase. Therefore the high temperature behaviour depends primarily on the composition, distribution, and crystallinity of the intergranular phase. The additive systems containing different rare-earth sesquioxides have been proved to improve the high temperature performance of the intergranular phase in the case of Si_3N_4 and SiAlON materials. Koushik Biswas[25] studied systematically the effects of different pairs of rare-earth sesquioxides and AlN-rare-earth oxide mixtures on the sinterability of SiC. The results indicated that sintering additives consisting of a single rare earth component are found to be unsuitable for the production of dense SiC due to insufficient liquid phase formation together with the in-situ SiO_2. β-SiC, seeded with α-SiC, was successfully sintered to full density with Gd_2O_3-Ho_2O_3 and Dy_2O_3-Ho_2O_3 additives. Dense SiC ceramics were also obtained with Lu_2O_3 substituted for Y_2O_3 in the "conventional" Y_2O_3-AlN additive system, and Lu_2O_3-AlN proved to be a promising additive system for significantly improved high temperature properties of LPS-SiC. More than 99% densification was achieved in all of these systems when sintering was carried out in N_2 atmosphere. N_2 is more effective for full densification as compared to argon atmosphere.

We have produced the SiC products by pressureless sintering process with average grain size 0.5μm SiC powder as raw material, using 1% (mass fraction) B_4C and 7% (mass fraction) C as sintering aids. The properties of the products are as following: density 3.10g/cm^3, fracture strength 320MPa, fracture toughness 4.21MPa · $m^{1/2}$, Vickers hardness 25.3 GPa, thermal conductivity 110W/(m · K).

4.2.2.2 Pressureless sintering of boron carbide ceramics

Due to the very strong covalent bonds, the high resistance to grain boundary sliding and absence of plasticity, densification of stoichiometric boron carbide (B_4C) is extremely difficult. In order to obtain full dense boron carbide ceramics, usually hot pressing technique is industrially useful. But for preparation of boron carbide ceramics with complex shape, pressureless sintering is often preferable. During the last decade, many investigators have carried out much research work related to the pressureless sintering of B_4C ceramics with different additives.

The best known additive for pressureless sintering of B_4C is carbon. Carbon additions have usually been through in situ-pyrolysis of phenolic resin which serves as a pressing agent and later yields 50% (mass fraction) carbon on decomposition. During the sintering, boron oxide reacts with carbon removed the B_2O_3 coatings separating B_4C particles from mutual contact. Schwetz and Vogt[26] first used a phenolic resin to synthesis boron carbide powders with very fine, pure, stoichiometric B_4C having a 10 ~ 22m^2/g surface area. They observed that the added carbon facilitated densification to 98% of theoretical density after heat-treating at 2150℃. Suzuki et al[27] sintered B_4C with 25% ~ 30% (mass fraction) carbon additions up to 90% ~ 93% of theoretical density after heat-treating at 2250℃, and observed the formation of a liquid of the eutectic composition (30% C). Schwetz and Grellner[28] studied the effect of carbon additions on the microstructure, and mechanical properties of sintered boron carbide in detail. They found that only the use of B_4C starting powder with a high surface area (above $15m^2/g$) along with carbon additions could produce dense (greater than 95% of T. D.) articles with a fine grained microstructure. Hyukjae Lee and Robert F. Speyer[29] studied the densification behaviors of pressureless sintering of B_4C with 3% (mass fraction) carbon doping in the form of phenolic resin. They found that densification started at 1800℃, the rate of densification increased rapidly in the range 1870 ~ 2010℃, which was attributed to direct B_4C-B_4C contact between particles permitted via volatilization of B_2O_3 particle coatings. Limited particle coarsening, attributed to the presence or evolution of the oxide coatings, occurred in the range 1870 ~ 1950℃. In the temperature range 2010 ~ 2140℃, densification continued at a slower rate while particles simultaneously coarsened by evaporation-condensation of B_4C. Above 2140℃, rapid densification ensued, which was interpreted to be the result of the formation of a eutectic grain boundary liquid, or activated sintering facilitated by nonstoichiometric volatilization of B_4C, leaving carbon behind. M. Bougoin and F. Thevenot[30] reported a result about the pressureless sintering of boron carbide with an addition of polycarbosilane and small amount of phenolic resin. The results dedicated that it is possible to obtain B_xC ceramics with high relative density ($\geqslant$92% T. D.) and without free carbon. It is necessary to introduce a minimal amount of carbon which could act in deoxidizing the boron carbide and to create a second phase which inhibits grain coarsening.

Some researchers have studied the pressureless sintering of B_4C ceramics with metal, oxide and boride additions. Mehri Mashhadi et al[31] analyzed the effect of Al addition on pressureless sintering of B_4C. The results indicated that addition of Al to B_4C samples increases the density and the grain size and decreases the porosity of sintered samples at various sintering temperatures; during the sintering, B_4C and Al react and useful phases of Al_3BC and AlB_2 are formed. By adding Al even 1% (mass fraction) and sintering at 2150℃ the shrinkage of sintering will be considerably. Density and grain size of sintered samples, increased significantly with Al load while less evidence is the effect of sintering temperature; 94% dense material was obtained by adding 4% (mass fraction). Bending strength, hardness and fracture toughness of sintered B_4C samples were shown to increase for Al content up to 4% (mass fraction) while further additions resulted in a decrease of the mechanical resistance. Conversely, elastic modulus showed an increase with Al load especially between 1% and 3% (mass fraction). Takeshi Kumazawa et al[32] studied the densification of boron carbide by pressureless sintering at temperature ranging from 2108 ~ 2226℃ with high purity B_4C, aluminum powder and SiC powder as starting materials. They obtained B_4C ceramics with a nearly full bulk density of 2.455g/cm^3 after being sintered at 2226℃. N. Frage et al[33] studied the effect of Fe addition on the densification of B_4C powder by spark plasma sintering. The results indicated that fully dense boron carbide can be produced from low-cost B_4C powder with 3.5% ~ 5.5% (volume fraction). Fe at relatively low temperature 2000℃ and short-time durations using the SPS approach; the formation of a transient liquid Fe-B-C solution increases the rate of sintering over the 1150 ~ 1750℃ temperature range.

Adrian Goldstein et al[34] studied the reactions occurring from room temperature to 2180℃ during the heating under argon of mixtures of B_4C and metal oxides, as well as the properties of the B_4C/metal boride composites. The oxides investigated included TiO_2, ZrO_2, V_2O_5, Cr_2O_3, Y_2O_3 and La_2O_3. Their results showed that the metal borides and B_4C are the majority phases, and the boron carbide/metal boride(s) mixtures resulted from these reactions exhibited a sintering aptitude significantly higher than that of pure boron carbide. The improvement in the sintering aptitude was proportional to the oxide content present in the initial mixture, up to an upper limit. B_4C/boride (s)-type composites, exhibiting bulk densities no less than 97% T.D., could be prepared for certain compositions by pressureless heating at 2180℃. Specially, in the case of the B_4C/Y_2O_3 system, a 98.5% relative density can be reached starting from a composition which includes no more than 8% (volume fraction) of oxide. The hardness and strength of the composites are in the range of the values measured for hot pressed B_4C.

S. Yamada et al[35] fabricated the B_4C based ceramics composites with 0 ~ 25% (mole fraction) CrB_2 by pressureless sintering in the temperature range 1850℃ to 2030℃. Their results indicated that the CrB_2 addition enhanced the densification of B_4C due to the CrB_2-B_4C eutectic liquid phase formation. Both a high strength of 525 MPa and a modest fracture toughness of 3.7MPa·$m^{1/2}$ were obtained for the B_4C-20% (mole fraction) CrB_2 composite with a high relative density of 98.1% after sintering at 2030℃. The improvement in fracture toughness is thought to result from the formation of microcracks and the deflection of propagating cracks resulting from the thermal expansion mismatch of CrB_2 and B_4C.

Hamid Reza Baharvandi et al[36] studied the effect of TiB_2 addition on sintering behavior and mechanical properties of pressureless-sintered B_4C ceramic. Different amounts of TiB_2, mainly 5% to 30% (mass fraction) were added to the base material. Pressureless sintering was conducted at 2050℃ and 2150℃. Addition of 30% (mass fraction) TiB_2 and sintering at 2150℃ resulted in improving the density of the samples to about 99% of T. D. The composite samples exhibited very good mechanical properties. As the amount of TiB_2 was increased further, the mechanical properties were reduced, except for the fracture toughness, apparently due to too much TiB_2 in the specimen.

We reported a result of pressureless sintering B_4C ceramics with rare-earth oxides as sintering aid[37]. Rare-earth oxides have been reported to be excellent sintering aids in ceramics. On the other hand, some rare earth elements, too, are characterized by a huge neutron absorption cross section (e. g. Gadolinium 46000, Samarium 5600, Europium 4300, and Dysprosium 950). It is the main reason using rare-earth compounds as sintering aids to promote the densification and to enhance the neutron absorption properties of B_4C ceramics. The results indicated that rare-earth additions are found to be excellent in lowering the sintering temperature. The relative density of specimen with 4% (mass fraction) Dy_2O_3 and 18% (mass fraction) phenolic resin sintering at 2040℃ is 96.6%, which achieves the similarly high relative density of specimen sintering at 2160℃ with only 18% (mass fraction) phenolic resin addition. The relative density of specimens with 4% (mass fraction) Eu_2O_3 and 4% (mass fraction) Sm_2O_3 first increase with the sintering temperature, the optimal sintering temperature is 2080℃ and 2040℃, with the relative density of 96.3% and 96.0% respectively, which is also higher than that of specimen without rare-earth, showing that the addition of Eu_2O_3 and Sm_2O_3 is also beneficial for a higher densification rates. The flexural strength of B_4C matrix ceramics is increasing obviously with the addition of rare-earth oxides. For specimen with addition of 4% (mass fraction) rare-earth oxides and 18% (mass fraction) phenolic resin, the flexural strength of more than 300MPa were obtained after sintering at 1960~2080℃. Figure 4.28 shows the SEM images of fracture surface of specimens B_4C with 18% (mass fraction) phenolic resin (a) and B_4C with 4% (mass fraction) of Dy_2O_3 and 18% (mass fraction) phenolic resin (b). The fracture surface of monolithic B_4C was quite smooth (Figure 4.28 a), with a few pores dispersed in the matrix. However, the fracture surface of specimen with 4% (mass fraction) of Dy_2O_3 (Figure 4.28 b) was relative rough and Dy_2O_3 particles were dispersed in intergranular areas. It is also apparent that the grain size in these samples is smaller than that of the rare-earth additive-free samples, which indicates that the addition of rare-earth prohibit the grain growth of B_4C particles, resulting in enhanced mechanical properties of B_4C matrix ceramics.

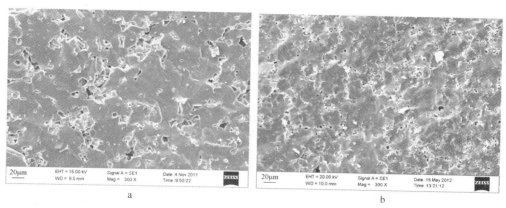

Figure 4.28 Microstructures of specimen sintered at 2040℃ for 2h
a—B_4C with 18% phenolic resin; b—B_4C with 4% of Dy_2O_3 and 18% phenolic resin

4.2.3 Hot Pressing Sintering

Hot pressing is often referred to as pressure sintering. It is a sintering method that pressure and temperature are applied simultaneously. The application of pressure at the sintering temperature accelerates the kinetics of densification by increasing the contact stress between particles and by rearranging particle positions to improve packing. It has been established that the energy available for densification is increased by greater than a factor of 20 by the application of pressure during sintering. Hot pressing has following advantages:

(1) Reduces densification time.

(2) Can reduce densification temperature, often resulting in less grain growth than that would occur with pressureless sintering.

(3) Minimizes residual porosity.

(4) Results in higher strength than can be achieved through pressureless sintering, due to the minimization of porosity and grain growth.

(5) Can reduce the amount of sintering aid and result in improved high-temperature properties.

(6) Can be conducted starting with a loose powder so that no binders or other organ-

ic additives are required.

Keys to hot pressing are equipment design and die design. Figure 4.29 shows a simple schematic of a typical uniaxial hot-pressing setup. It consists of a furnace surrounding a high-temperature die with a press in-line to apply a controlled load through the die pistons. The type of furnace is dependent on the maximum temperature and uniformity of the hot zone required. Induction heating, with water-cooled copper coils and a graphite susceptor, is most commonly used and has a temperature capability greater than 2000℃. The furnace must either be evacuated or backfilled with N_2, He or Ar during operation to minimize oxidation of the graphite. Furnaces with graphite or other resistance heating elements can also be used for hot pressing. Figure 4.30 is a photo of hot pressing furnace with graphite as heat element.

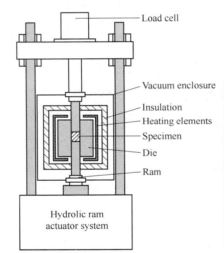

Figure 4.29 Schematic of the hot pressing process

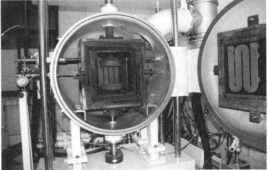

Figure 4.30 Photo of hot pressing furnace

The source of pressure is usually a hydraulic press with a water-cooled platen attached to the ram. However, this does not provide adequate cooling to extend the ram into the furnace, so blocks of graphite or other refractory material are used. Obviously, the size of the press is dependent on the size of the part being hot pressed and the pressure required.

The die material is perhaps the most important element of the hot press. It must withstand the temperature, transient thermal stresses, high hot-pressing loads, and be chemically inert to the material being hot pressed. Graphite is the most widely used die and piston material. It has high-temperature capability, its strength increases with temperature, and it has low coefficient of friction. It does not react with most materials and can be coated with a boundary layer such as boron nitride (BN) to prevent direct contact with material with which it might interact. As with the graphite susceptor, though, graphite does oxidize and must be used under a protective environment.

Refractory metal dies such as molybdenum, tantalum, and the molybdenum alloy TZM (Titnaium-Zirconium-Molybdenum Alloy) have been used in limited cases. However, they are expensive, have high reactivity, and deform easily at high temperatures. TZM coated with $MoSi_2$, or a composite die consisting of a molybdenum jacket surrounding an Al_2O_3 liner has been recommended. This latter approach takes advantage of the strength of the molybdenum and the abrasion resistance, creep resistance, and moderate thermal expansion coefficient of the Al_2O_3. Superalloys have also been used for hot-pressing dies for ceramics, but only at temperatures below 900℃ and loads below 104 MPa. A major problem with these materials is high thermal expansion. If the expansion of the die is higher than that of the material being hot pressed, the die will essentially shrink fit around the material during cooling and make ejection extremely difficult.

Ceramic dies, especially Al_2O_3 and SiC, have been used successfully for hot pressing. They have reasonably low thermal expansion, are non-reactive, and have excellent resistance to galling and abrasion. Al_2O_3 can be used to approximately 1200℃, dense SiC to about 1400℃.

Reactivity is a special concern of die assemblies. Many of the carbides, ferrites, and other materials are very susceptible to property alteration through variations in stoichiometry and must be hot pressed under very controlled conditions. Graphite dies are often lined with a "wash" or spray coating of BN or Al_2O_3. Dies used for ferrites and some other electronic ceramics are often lined with ZrO_2 or Al_2O_3 powder.

In our laboratory, we apply the C/C die to hot pressing, the pressure can reach 100MPa at 2200℃.

The nature of the powder to be hot pressed is equally important to correct selection of the die material. The same type of fine-grained powders suitable for pressureless sintering is usually acceptable for hot pressing. In most cases a densification aid or a grain-growth inhibitor is added to achieve maximum density and minimum grain size.

Hot pressing is typically conducted at approximately half the absolute melting temperature of the material, which is usually a lower temperature than that at which the material can be densified by pressureless sintering. Time at temperature is also reduced. The reduced temperature and time at temperature combine to minimize grain growth, thus providing better potential for improved strength.

Powder to be hot pressed can be loaded directly into the die or can be precompacted separately into a powder preform or compact that is then loaded into the die. Loading powder directly into the die is the most common procedure. However, the problems with this method are the difficulty in achieving uniformity and the pickup of contamination. Another disadvantage involves the low packing density of the loose powder and the resulting increase in the stack height to achieve a given part thickness. This reduces the number of parts that can be produced in a hot-pressiug run and also increases the die-wall friction. Increased die-wall friction increases the variation of pressure within the compact and increases the chances for nonuniformity in the final part.

Hot pressing permits achieving near-theoretical density and very fine grain structure, which result in optimization of strength. It also permits reduction of the amount of sintering aid required to obtain full density. This can result in orders-of-magnitude improvement in high-temperature properties such as creep and stress rupture life.

Table 4.7 compares the properties of several sintered and hot-pressed Si_3N_4 compositions. Similar differences exist between sintered and hot-pressed varieties of other materials such as Al_2O_3, SiC, spinel, and mullite.

Table 4.7 Comparison of densities and strengths achieved by hot pressing vs. sintering

Material	Sintering aid	Density (theoretical) /%	Strength (RT) /MPa	Strength (1350℃) /MPa
Hot-pressed Si_3N_4	5% MgO	98	587	173
Sintered Si_3N_4	5% MgO	about 90	483	138
Hot-pressed Si_3N_4	1% MgO	>99	952	414
Sintered Si_3N_4	BeSiN-SiO_2	>99	560	—
Sintered Si_3N_4	6% Y_2O_3	about 98	587	414
Hot-pressed Si_3N_4	13% Y_2O_3	>99	897	669

Hot pressing can cause preferred orientation of the grain structure of some materials and result in different properties in different directions. This occurs predominantly when powders with a large aspect ratio such as rods or needles are used. It can also occur due to flattening of agglomerates or laminar distribution of porosity perpendicular to the direction of hot pressing.

Preferred orientation has also been encountered in hot-pressed Al_2O_3 reinforced with SiC whiskers. The whiskers have a high length-to-diameter ratio (usually over 20∶1)

and orient perpendicular to the hot-pressing direction. Test bars cut from a plane perpendicular to hot pressing break across the whiskers and have high strength (>600MPa) and toughness (7MPa $\cdot$ m$^{1/2}$). Bars cut from a plane parallel to the hot-pressing direction break parallel to the whiskers and have much lower strength (typically <400MPa) and toughness (3.5 to 4.0MPa $\cdot$ m$^{1/2}$). Strength test specimens are normally cut from the plane perpendicular to the hot-pressing direction. This is usually the strongest direction (if anisotropy is present).

The major limitation of hot pressing is shape capability. Flat plates, blocks, or cylinders are relatively easy to hot press. Long cylinders, nonuniform cross sections, and intricate shapes are difficult and often impossible by conventional uniaxial techniques.

The starting powder goes into the die as a relatively uniform stack of powder or as a uniform preform. During densification the powder or preform will compact in the axial direction of pressure application until the porosity has been eliminated and near-theoretical density achieved.

4.2.4 Hot Isostatic Pressing

Most of the limitations of hot pressing result from the uniaxial pressure application. Techniques have been developed to hot press from multiple directions. This is referred to as hot isostatic pressing (HIP) and is analogous to cold isostatic pressing. Apparatus for HIP consists of a high-temperature furnace enclosed in a water-cooled autoclave capable of withstanding internal gas pressures up to about 320GPa and providing a uniform hot-zone temperature up to about 2000℃. Pressurization gas is either argon or helium. Heating is usually by molybdenum or graphite resistance-heated elements. A schematic of a HIP apparatus is shown in Figure 4.31.

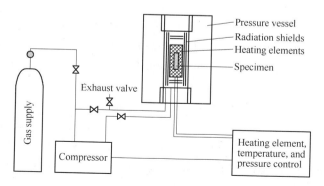

Figure 4.31 Schematic of hot isostatic pressing equipment[38]

To achieve densification of a ceramic preform, the preform must first be evacuated and then sealed in a gas-impermeable envelope. If any high-pressure gas leaks into the preform, pressure is equalized and the preform cannot hot press. The earliest HIP studies encapsulated the ceramic in tantalum or other metal, depending on the temperature required for densification. However, this severely limited the shape capability. Later studies were directed toward the use of glass encapsulation. The glass was applied as a preformed envelope, which was sealed around the part under vacuum and then collapsed at high temperature to conform to the ceramic preform shape. This worked for relatively simple shapes.

HIP has the potential of resolving some of the major limitations of uniaxial hot pressing. It makes possible net-shape forming because the pressure is equally applied from all directions. This also results in greater material uniformity by eliminating die-wall friction effects and preferred orientation, resulting in higher strength and Weibull modulus. Also, much higher pressures and temperatures can be used, making possible more complete densification and greater flexibility in selection of composition. For instance, the higher pressure and temperature may permit densification of compositions containing less sintering aid and having dramatically improved stress rupture life and oxidation resistance.

4.2.5 Gas Pressing Sintering

Hot isostatic pressing of powder compacts is usually done after encapsulation of the compacts within a container, using an inert gas of a higher pressure than that of HP, up to about 300MPa. As a variation of HIP, gas pressure sintering (GPS) usually under 10MPa is available. GPS makes possible the sintering of materials with a high vapour pressure, such as Si_3N_4.

GPS process is one of the most promising sintering techniques for the fabrication of high performance silicon nitride ceramics. Almost pore-free densified compacts with low amounts of sintering additives can be obtained by this process. They present high strength with high reliability and good heat resistance. These properties make them one of the best candidate materials for heat engine component as well as other industrial applications. The gas pressure sintered silicon nitride (GPSSN) ceramics densified to high density have already been applied in turbocharger rotors for automotive application, cutting tools and bearing balls for industrial applications, etc, and have been highly evaluated for high strength and reliability.

In gas pressure sintering, the pressure, the temperature and the sintering time, are closely related. With a higher temperature, it is necessary to decrease a sintering time, but in order to avoid the decomposition it is required a higher gas pressure. Although the simultaneous increase of temperature and N_2-pressure make it possible to obtain higher densities, full densification can never be achieved by one step pressure sintering. Densification is hindered by the inclusion of gas under pressure in the porosity. The use of higher sintering temperatures also decreases the fracture toughness due to the grains coarsening and globularization of the rod-like β-Si_3N_4 grains.

In GPS, two-stage sintering is usually performed: low pressure sintering before the isolation of pores and high pressure sintering after the pore isolation.

A two-stage nitrogen pressurizing technique with pressures of up to 7MPa being applied after de closed porosity stage is reached, has achieved densities of above 99%[39]. With this technique it is possible that the production of complex-shaped components with properties is similar to those of hot pressed material. This method allows the sintering of complex-shaped parts with medium cost. The advantage of this technique over pressureless sintering is that it allows high sintering temperature for long periods of time without decomposition of the ceramic material. Therefore, the high gas pressure prevents the decomposition (sublimation) of the silicon nitride.

The experience has shown that a sintering schedule with low temperature and low pressure for a relatively long period of time, followed by a step-up in both temperature and pressure, gave the densest samples with the lowest mass loss (decomposition). This behavior is probably due to the formation of an outer region with closed porosity that forms a dense shell. Therefore, the final step-up pressure results in an isostatic pressure on the inner, porous bulk material that closes the pores and speeds up the densification.

Different sintering aids and aids compositions have a great influence on the densification of silicon nitride.

On the other hand, for the Si_3N_4 ceramics of high temperature microwave transmission application, they do not need full dense, one step pressure sintering is effective. We studied the effects of sintering aids (Y_2O_3, YAG and dry gel of YAG) on mechanical and dielectric properties of Si_3N_4 ceramics sintered at 1800℃ under 8MPa N_2 atmosphere. The results show that the sintering aids have a significant impact on the densification and mechanical properties of Si_3N_4 ceramics. Of all, the Si_3N_4 ceramics using the dry gel of yttrium aluminium garnet as sintering aids exhibit the maximum bending strength and fracture toughness, reaching 395MPa and 4.25MPa $\cdot$ $m^{1/2}$ respectively. However, the dielectric properties of Si_3N_4 ceramics are similar to that using the other two sintering aids. The improved mechanical properties of the Si_3N_4 ceramics are attributed to the formation of rod-like β-Si_3N_4 particles[40].

4.2.6 Reaction Sintering

Reaction sintering, or reaction bonding, is an important means of producing dense covalent ceramics. This generally involves mixing the reactants together as powders, compacting the powders into the desired shape, and heating to a temperature at which the powders react with gas or liquid reactants to form ceramic materials in the furnace.

Two important examples are reaction-sintered Si_3N_4 and reaction-sintered SiC.

Reaction-sintered Si_3N_4 is also referred to as reaction-bonded Si_3N_4 or simply RBSN. RBSN is fabricated from silicon powder. The silicon powder is processed to the desired particle size distribution and formed into the required shape by pressing, slip casting, injection molding, or another suitable process. The compacted Si shape is then placed in a furnace under a nitrogen or mixed nitrogen and hydrogen or nitrogen and helium atmosphere and heated initially to about 1200 to 1250℃. The nitrogen permeates the porous Si compact and begins to react with the Si to form Si_3N_4. Initially, α-Si_3N_4 fibers grow from the Si particles into the pores. As the reaction progresses, the temperature is slowly raised to approximately 1400℃, near the melting temperature of Si. As the temperature increases, the reaction rate increases and primarily β-Si_3N_4 is formed. Great care is necessary in controlling the rate of temperature increase and nitrogen flow. The reaction of N_2 and Si is exothermic and, if allowed to proceed too fast, will cause the silicon to melt and ball up into Si particles too large to nitride or which exude out of the surface of the part. A typical nitriding cycle in which the exotherm (heat rise and fall vs time) is controlled and no exuding occurs is on the order of 7 to 12 days, depending on the volume of material in the furnace and the green density of the starting Si compacts.

Approximately 60% weight gain occurs during nitriding, but less than 0.1% dimensional change. This makes possible excellent dimensional control. Bulk densities up to 2.8g/cm^3 have been achieved (compared to a T.D. for Si_3N_4 of about 3.2g/cm^3).

The earliest reaction-bonded Si_3N_4 had a density of about 2.2g/cm^3 and a strength under 138MPa. By the late 1970s, RBSN had been developed with density of 2.8g/cm^3 and four-point flexure strength in the range of 345MPa.

Another advantage of the reaction-sintered Si_3N_4 is its creep resistance. No sintering aids are added to achieve densification, so no glassy grain boundary phases are present.

Strength is retained to temperatures greater than 1400℃ and the creep rate is very low. In addition, reaction sintered Si_3N_4 has a relatively low elastic modulus and coefficient of thermal expansion and a relatively high thermal conductivity (considering its porosity). These properties, combined with a moderately high strength, give RBSN good thermal shock resistance and make it a feasible candidate for such applications as welding nozzle tips and some prototypes gas-turbine static-structure components.

The primary disadvantage of RBSN is its porosity. The porosity is interconnected and can result in internal oxidation and accelerated surface oxidation at high temperature. The internal oxidation appears to affect the thermal stability of a component. The mechanism could have been associated with internal stresses induced by the thermal expansion mismatch between the Si_3N_4 and the cristobalite formed during oxidation. To avoid oxidation, a post-sintered RBSN was developed. A sintering aid such as Y_2O_3 was added to the Si prior to nitriding. After nitriding, the 80%-dense RBSN was heated to around 1900℃ in nitrogen (preferably with an overpressure) and sintered by a liquid-phase mechanism to near-theoretical density (typically less than 1% porosity). This material had a strength greater than 700MPa and very good oxidation resistance.

Reaction-bonded silicon carbide (RBSC) is also referred to as reaction-bonded SiC and siliconized SiC. It is processed from an intimate mixture of SiC powder and carbon. Bodies formed from this mixture are exposed to liquid or vapour silicon at 1500 ~ 1700℃ high temperature. The silicon reacts with the carbon to form new silicon carbide, which bonds the original particles together. Silicon also fills any residual open pores. Figure 4.32 shows the main elements of RBSiC; α-SiC from the starting powder, β-SiC formed during the reaction process and remaining Si.

Shandong Baona New Materials Ltd. produces the reaction-sintered SiC products, with the composition of SiC powder A 55% (mass fraction), SiC powder B 35% (mass fraction), lamp black and graphite powder 9% (mass fraction), dispersant 0.7% (mass fraction), polyvinyl alcohol (PVA) 0.3% (mass fraction), through mixing the mixture of raw materials to prepare slurry, slip casting then sintering at 1720℃ in a vacuum furnace. During the loading, a content of Si powder is added with a ratio of green body weight : Si powder weight of 1 : 0.72. The properties of product are following: fracture strength 326MPa, fracture toughness 4.47MPa·$m^{1/2}$, Viker hardness 25GPa, thermal conductivity 80W/(m·K), density 3.09g/cm^3, free C 7.76% (mass fraction). The SEM image of fracture surface of the sample is shown in Figure 4.33. The practical sintering curve is shown in Figure 4.34.

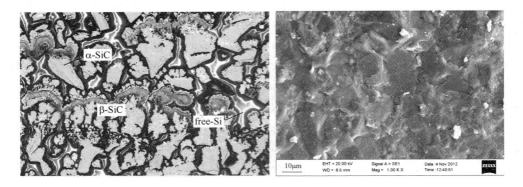

Figure 4.32 The grain structure of RBSiC

Figure 4.33 SEM image of RBSiC sintered at 1720℃

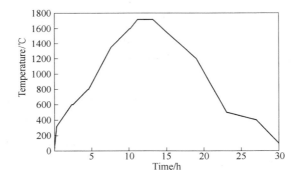

Figure 4.34 The sintering curve of reaction-bonded SiC

Reaction-sintered SiC has similar advantages to reaction-bonded Si_3N_4 for complex-shape fabrication; it undergoes dimensional change of less than 1% during densification. The initial shape can be formed by casting, plastic molding, pressing, extrusion, and any of the other processes applicable to ceramics. It is especially suitable for the plastic processes such as extrusion and compression molding. The binder can be a thermosetting resin such as a phenolic. Instead of having to remove the binder after molding, as is required with other ceramics, the binder is simply charred to provide the carbon source for reaction with the silicon.

The main disadvantage is inferior heat resistance over 1300℃, the fracture strength decreases as the temperature increasing in the range of high temperature, due to the existing of free Si. The corrosion, abrasion and heat resistance will vary depending upon the material's content and distribution of free silicon, grain size, and free carbon content.

Another interesting method of fabricating reaction-sintered SiC is to start with woven carbon fibers or felt. Laminae of carbon fiber weave are laid up much the same as fiberglass to form the desired shape. This can then be reacted with molten silicon to form a composite of SiC-Si or C-SiC-Si. By controlling the type of fibers and the weave, a complete range of composites with varying SiC-to-Si ratios can be engineered. Those with high Si have low elastic modulus (about 206GPa) and relatively low strength (< 206MPa); those with high SiC have high elastic modulus (> 344GPa) and high strength (about 482MPa). The range of composites can be further expanded by pitch impregnation or deposition of pyrolytic carbon or glassy carbon prior to siliconizing.

4.2.7 Spark Plasma Sintering

Spark plasma sintering (SPS) is a field-assisted compaction method which allows very rapid heating (up to 600℃/min or more) and cooling rates, very short holding times (minutes), and the preparation of fully dense samples at comparatively low sintering temperatures, typically a few hundred degrees lower than in normal hot pressing. Instead of using an external heat source (as in conventional hot pressing), an electrical current (DC, pulsed DC, or AC) is allowed to pass through the conducting pressure die and, in some cases, also through the sample; in this way the die itself acts as a heat source, so the sample is heated from both inside and outside.

A spark sintering method was investigated and patented in the 1960s and used to compact metal powders, but due to high equipment cost and low sintering efficiency it was not put to wider use. The concept was further developed during the mid 1980s to the early 1990s, and a new generation of sintering apparatus appeared named Plasma Activated Sintering (abbreviated PAS) and Spark Plasma Sintering (abbreviated SPS). Common to these systems is the use of pulsed direct current to heat the specimens. These sintering techniques currently attract growing attention among productions engineers as well as materials researchers. Whether plasma is generated has not been confirmed yet, especially when non-conduction ceramic powders are compacted. It has, however, been experimentally verified that densification is enhanced by the use of a pulsed DC current or field[41]. This family of techniques is in academia also named as pulsed electric current sintering (PECS)[42] or electric pulse assisted consolidation (EPAC)[43].

Figure 4.35 shows a basic configuration of a SPS unit. It consists of a uniaxial pressure device, where the water-cooled punches also serve as electrodes, a water-cooled reaction chamber that can be evacuated, a pulsed DC generator, pressure-regulating, position-regulating and temperature-regulating systems. Hong peng[44] gave a graph of ON-OFF pulsed direct current during the entire sintering cycle, with a standard on-off time relation of 12 : 2 in his doctoral dissertation. Spark plasma sintering resembles the hot pressing process in several respects, i. e. the precursor powder (green body) is loaded in a die, and a uniaxial pressure is applied during sintering process. However, instead of using an external heating source, a pulsed direct current is allowed to pass through the electrically conducting pressure die and, in appropriate cases, also through the sample. This implies that the die also acts as a heating source and that the sample is heated from both outside and inside. The use of a pulsed direct current also implies that the samples are exposed to a pulsed electric field during the sintering process.

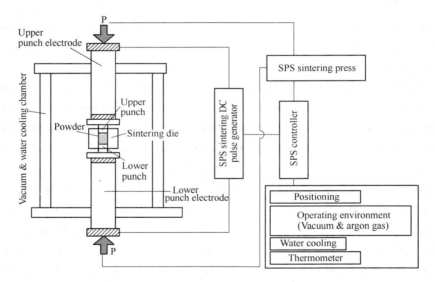

Figure 4.35 Basic configuration of a typical SPS system[45]

The ON-OFF DC pulse energizing method generates: (1) spark plasma; (2) spark impact pressure; (3) Joule heating; (4) an electrical field diffusion effect. In the SPS process, the powder particle surfaces are more easily purified and activated than in conventional electrical sintering processes and material transfers at both the micro and macro levels are promoted, so a high-quality sintered compact is obtained at a lower temperature and in a shorter time than with conventional processes.

Conventional electrical hot press processes use DC or commercial AC power, and the main factors promoting sintering in these processes are the Joule heat generated by the power supply (I^2R) and the plastic flow of materials due to the application of pressure. The SPS process is an electrical sintering technique which applies an ON-OFF DC pulse voltage and current from a special pulse generator to a powder of particles, and in addition to the factors promoting sintering described above, also effectively discharges between particles of powder occurring at the initial stage of the pulse energizing for sintering. High temperature sputtering phenomenon generated by spark plasma and spark impact pressure eliminates adsorptive gas and impurities existing on the surface of the powder particles. The action of the electrical field causes high-speed diffusion due to the high-speed migration of ions.

When a spark discharge appears in a gap or at the contact point between the particles of a material, a local high temperature-state (discharge column) of several to ten thousands of degrees centigrade is generated momentarily. This causes evaporation and melting on the surface of powder particles in the SPS process, and "necks" are formed around the area of contact between particles.

It is necessary to plan the sintering parameters that yield fully dense compacts with a very limited grain growth. The critical temperature above which the grain growth rate becomes appreciable is largely determined by the properties of the precursor powders, e.g. their particle size, reactivity, degree of agglomeration etc, but also by the applied heating rate and pressure.

The common phenomena so far observed during the SPS processing of ceramics can be summarized as follows:

(1) Densification and grain growth processes are accelerated and take place sequentially, providing us with the possibility to separate the grain growth and densification processes that in turn allows us to monitor the kinetics of these processes.

(2) The microstructures of SPS consolidated materials can be precisely regulated, due to better understanding and control of the sintering and grain growth kinetics.

The use of rapid heating rates appears to be essential in order to obtain fully dense nano-sized functional ceramics. The use of rapid heating rates implies that the sample is exposed to thermal gradients that in turn promote the kinetics of grain growth and densification. The use of rapid heating rates also implies that the grain growth at the early stage of sintering is suppressed and that the formation of undesirable intermediate phases can be avoided.

The possibility of applying high mechanical pressure in SPS process increases the possibility of manufacturing hard-to-make materials, e.g. of obtaining fully dense ceramics within shorter time and at lower temperature than through a conventional hot-pressing process.

Spark plasma sintering (SPS) is a high-speed powder consolidation/sintering technology capable of processing conductive and nonconductive materials. Theories on the SPS process vary, but most commonly accepted is the micro-spark/plasma concept, which is based on the electrical spark discharge phenomenon wherein a high-energy, low-voltage pulse current momentarily generates spark plasma at high temperatures (many thousands of ℃) in fine local areas between particles. SPS' operational or "monitored" temperatures (200 ~ 2400℃) are commonly 200 to 500℃ lower than with conventional sintering, classifying SPS as a lower-temperature sintering technology. Material processing (pressure and temperature rise and hold time) is completed in short periods of approximately 5 to 25 minutes. The relatively low temperatures combined with fast processing times ensure tight control over grain growth and microstructure.

SPS is a very powerful technique not only to densify ceramics, but also to prepare many kinds of material. From all the examples found in the literature, Teresa Hungria et al[46] summarized three main applications of this sintering technique, i. e., compaction of ceramics, preparation of composites, and SPS in situ reaction sintering.

SPS has been demonstrated to enable the consolidation of ceramics materials within minutes and the ability to obtain fully dense samples at comparatively low sintering temperatures, typically a few hundred degrees lower than in normal hot pressing. The most commonly compacted materials by this technique are: carbides, for example, SiC[47], Ti_3SiC_2, B_4C[48~51], WC and ZrC; nitrides, for example Si_3N_4, ZrN, TiN[52], and AlN[53,54]; oxides, for example, ZrO_2, TiO_2[55], Al_2O_3[56~58], MgO[59], Y_2O_3[60] and spinel[61]; phosphates, for example, hydroxyapatite; and borides, for example, ZrB_2 ultra-high-temperature ceramics[62].

In situ processing of composites is, compared to conventional processing, advantageous for obtaining materials with a finer and more homogeneous microstructure, high chemical and thermodynamical stabilities at a higher temperature, and better mechanical properties. SPS can be used to process a wide range of materials because of its short sintering time, which avoids reactions between the material components. Some composites are described in the literature, for example, the high-density $(1-x)(Na_{0.5}K_{0.5})NbO_{3-x}LiTaO_3$ ceramics[63], Al_2O_3-Co composites[64], Ti_3SiC_2-Cu metal-ceramic composites[65], Si_3N_4/SiC composites[66], SiC nano-whisker/TiCN-based cermets[67],

In_2O_3-Sr_2RuErO_6 composite[68], silicon carbide whiskers (SiCw) reinforced nanocrystalline alumina[69], functionally graded B_4C-Al composite, pseudo eutectic TiB_2-B_4C composites, HfB_2-20% (mass fraction) SiC composites[70], ZrC-SiC ceramics[71]. Recently, a grapheme/SiC ceramic composite by spark plasma sintering is reported[72]. The grapheme was in-situ grown within either α-phase or β-phase SiC ceramics during their densification by spark plasma sintering (SPS).

Some recent studies show the possibility of using the SPS technique not only to create a fully dense material, but also to synthesize the constituent phases of the ceramics. For example, it is possible to obtain TiN/Al_2O_3 nanocomposites using Ti, AlN and TiO_2 commercial powders as starting materials. This novel fabrication process for a TiN/Al_2O_3 nanocomposite basically involves molecular-level mixing of the TiN and Al_2O_3 during a reaction process, instead of conventional powder mixing. The as-prepared TiN/Al_2O_3 nanocomposite has a finer microstructure with a grain size below 400 nm, and exhibits better mechanical properties[73]. It is also possible to obtain submicrometer TiC/SiC composites by a rapid reactive sintering process through spark plasma sintering (SPS) technique using the carbon, titanium, and nanosized-SiC powders without any additive. The composite sintered in a relatively short time (8 min at 1480℃) achieved 97.9% of T. D., and the research results also shown that SPS technique was an attractive way to fabricate submicrostructure TiC/SiC composites[74]. Similar results were observed for a mixture of Ti, C and Si powders to obtain Ti_3SiC_2-SiC composites[75]. In the case of the refractory materials, as for example HfB_2, the simultaneous synthesis and sintering produced dense ceramics (with 98% T. D.) at lower pressure and temperature[76].

One of the most interesting application of the SPS technique is the fabrication of nanostructured materials, when nanosized powdered precursors are used. This clear advantage over conventionally employed sintering methods can be attributed to the lower sintering temperature and shorter holding times needed, which allows nearly complete densification with little grain growth to be achieved. M. Nygren reported[77] SPS processing of nano-structured ceramics of $Bi_4Ti_3O_{12}$, $BaTiO_3$, and $SrTiO_3$, and discussed the effects of grain size of the starting powder on the grain growth factor. The grain growth factor varies with the grain size of the starting powder, i. e. if the starting powder has a grain size in the range 60 ~ 100nm the obtained compacts normally have grain sizes in the range of 150 ~ 250nm. With decreasing grain size of the starting powder, e. g. 10 ~ 30nm, the grain growth factor is increasing from 1.5 ~ 2.5 to 4 ~ 6. These observations are in line with the common knowledge that the driving force for reduction of the interfacial area is much smaller for 100 nm sized precursor powders than for 10 nm ones.

As mentioned above, the most important application of SPS technique is to fabricate nano-ceramic and nano-composites. In the last decades, many researchers focused on the spark plasma sintering of nano-ceramic and nano-composites.

Summing the nanoceramic spark plasma sintering studies so far, there are several advantages that accrue from this process. First, the rapid heat transfer and resistance heating of particles significantly shorten the high temperature exposure of nanoparticles, which is essential for preserving the nanostructures. The rapid heating (100 ~ 1000℃/min) in spark plasma sintering enhances the densification process which usually occurs at high temperatures. Apart from very fast heating rate, spark plasma sintering also allows for fast cooling rate, very short holding times, and the possibility to obtain fully dense nanostructures at comparatively low sintering temperatures, typically a few hundred degrees centigrade lower than that in the conventional pressure sintering. The sintering pressure for spark plasma sintering is lower than that of hot pressing and close to that of hot isostatic pressing, typically (200 MPa, which might be limited by the uppermost strength of the best available graphite moulds. These combined advantages offer very limited grain growth. The grain size of spark plasma sintered compact is almost definitely smaller than that from other densification techniques. Other advantages of spark plasma sintering technique include less sintering additives. This mainly stems from the accelerated interparticle bonding by the electric pulses which make the reduction/elimination of additives possible. However, spark plasma sintering also has its limitations. Because of the rapid densification process and the existence of multiple sintering driving forces, it is difficult to decouple the diffusion mechanisms and trace the exact sintering processes. Spark plasma sintering fundamental study is clearly needed. Most efforts are focused on achieving fully dense nanostructures without in-depth discussion of atomic diffusion or deformation processes. Also, the vast majority of spark plasma sintering is based on simple shape and small size samples even though a continuous production of compacts of complex geometry and above 150mm diameter pieces has been demonstrated, at least partly for proof of concept and partly to demonstrate the untapped capabilities of spark plasma sintering. Spark plasma sintering is not applicable to every material as discussed before. Sintering efficiency depends on the electrical properties of the material being sintered. The required electrical current, current source, and current discharge rate are all system specific and have to be studied for each material. Another disadvantage is the expensive equipment involved, such as the sintering set-up and the special dies capable of imparting high current to the compact at elevated temperature and pressure. Also, temperature gradient is inevitable in the samples. Under certain conditions, the difference may reach 450℃ or more. This can result in differential densification and subsequent microstructure inhomogeneity[78].

4.2.8 Microwave Sintering

Studies on microwave processing of ceramics started in the mid 1960s by Tinga (Tinga & Voss, 1968; Tinga & Edwards, 1968) and, since then, interest in the use of microwaves for heating and sintering ceramics has grown steadily. In recent years, microwave heating has been widely employed in the sintering and joining of ceramics[79,80].

J. G. P. Binner and B. Vaidhyanathan[81] answered what microwave sintering of ceramic is. Microwave heating is a process whereby microwaves couple to materials, which absorb the electromagnetic energy volumetrically and transform it into heat. This differs from conventional methods in which heat is transferred between objects through the mechanisms of conduction, radiation and convection. Because the material itself generates the heat, heating is more volumetric and can be very rapid and selective. Thus, microwave sintering techniques allows for the application of high heating rates, markedly shortening the processing time. Microwave processing eliminates the need for spending energy to heat the walls of furnace or reactors, their massive components and heat carriers. Hence, the use of microwave processing methods significantly reduces energy consumption, particularly in high temperature processes. However, the advantages of using microwave energy in high-temperature processes are by no means limited to energy savings. In many cases, microwave processing can improve the product quality.

In the case of most ceramics, the microwave heating is mainly characterized by the dielectric loss of the material. However, there are other factors that also significantly contribute to the microwave heating, such as ionic conductivity degree of porosity, particle size, electrical conductivity, magnetic coupling, etc. The exact mechanism of microwave heating and sintering has not yet been well explained and understood.

The interaction of an electric field with a material may elicit several responses, and microwaves can be reflected, absorbed and/or transmitted by the material. In a conductor, electrons move freely in the material in response to the electric field, resulting in electric current. Unless the material is a superconductor, the flow of electrons will heat the material through resistive heating. However, microwaves will be largely reflected from metallic conductors, and therefore such conductors are not effectively heated by microwaves. In insulators, electrons do not flow freely, but electronic reorientation or distortions of induced or permanent dipoles can give rise to heating. Because microwaves generate rapidly changing electric fields, these dipoles change their orientations rapidly in response to the changing fields. If the electric field change takes place close to the natural frequency at which reorientation can occur, the maximum amount of energy is consumed, resulting in optimum heating. In microwave processing terminology, this event is described by the term "well-coupled" material.

Depending on their electrical and magnetic properties, materials can be divided into three categories according to their microwave absorption properties. Materials with a very low dielectric loss factor (ε'') allow microwaves to pass through with very little absorption and are said to be transparent to microwaves. Materials with an extremely high dielectric loss factor, i. e. metals, reflect microwaves and are said to be opaque. Materials with intermediate loss tangents will absorb microwaves.

Bulk metals are opaque to microwave and are good reflectors—this property is used in radar detection. However, powdered metals are very good absorbers of microwaves and heat up effectively, with heating rates as high as 100℃/min[82].

Ceramics with loss factors between the limits of $10^{-2} < \varepsilon'' < 5$ are good candidates for microwave heating. Ceramics with $\varepsilon'' < 10^{-2}$ would be difficult to heat, while those with $\varepsilon'' > 5$ would absorb most of the heating on the surface and not in the bulk.

During the exploitation of microwave heating, most of the research on material processing by microwaves is based on conventional low-frequency (2.45GHz) microwave applicators. However, such applicators do not couple microwave power efficiently to many ceramics at room temperature, and poor microwave absorption characteristics make initial heating difficult. Thermal instabilities may occur, which can lead to the phenomenon of **thermal runaway**; i. e., the specimen overheats catastrophically. The temperature gradients inherent in volumetric heating can lead to severe temperature non-uniformities, which, at high heating rates, may cause non-uniform properties and cracking.

In general, the microwave absorption of many ceramics can be increased by raising the temperature, adding absorbents (e. g., SiC, carbon, binders), altering their microstructure and defect structure by changing their form (e. g., bulk vs. powder), or by changing the frequency of incident radiation.

Increasing the temperature (with radiant heat) is a common method used by many researchers to couple microwaves with poorly absorbing (low-loss) materials. This is called **hybrid heating** (**MHH**) technique. Once a material is heated to its critical temperature, microwave absorption becomes sufficient to cause self-heating. This hybrid method can result in more uniform temperature gradients because the microwaves heat volumetrically, and the external heat source minimizes surface heat losses. Hybrid heating can be achieved by using either an independent heat source, such as a gas or electric furnace in combination with microwaves, or an external **susceptor** that couples with the microwaves. In the latter, the material is exposed simultaneously to radiant heat produced by the susceptor and to microwaves.

One of the most important characteristics associated with the use of microwave hybrid heating is the potential to achieve uniform heating throughout the cross-section of a material, as illustrated in Figure 4.36.

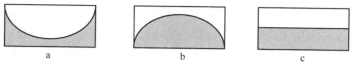

Figure 4.36 Temperature distribution in ceramic material heated by conventional fast firing (a), microwave fast firing (b) and microwave hybrid heating (c)[80]

An outstanding example of the use of microwave hybrid heating was the sintering of, $ZrO_2/8\% \ Y_2O_3$ and zirconia/12% CeO_2 at 2.45 GHz. Although these materials could be sintered readily at 28 GHz, attempts to use 2.45 GHz were unsuccessful until SiC rods were inserted into the insulation surrounding the specimens, in what is referred to as the "picket fence" arrangement (Figure 4.37). The microwave energy initially heated the SiC rods, which resulted in uniform heating of the zirconia, and hence, in a homogenous coupling with microwaves after the critical temperature was reached.

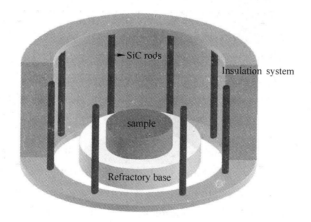

Figure 4.37 Illustration of the "picket fence" arrangement[83]

Another significant advantage associated with the use of microwaves is good "on-off" control, i. e. the product is only heated when the power is on and immediately it is turned off, the sample begins to cool. Thus microwave sintering is proving attractive for situations where great control over the microstructure is required. An example lies in the recent desire to create nanostructured ceramics. Although in general we are still a long way from creating the perfect green nanostructured ceramic (due to problems with particle agglomeration and difficulties in green forming the powders), in time these will be solved and a method of achieving densification whilst limiting grain growth will be required. The very fast heating rates and good "on-off" control potentially offer major advantages. Work to date has already shown that it is possible to limit grain growth whilst achieving significant levels of densification though the quality of the green body is critical. The lower the green density, the greater the degree of densification is required. This results in longer sintering periods and hence more opportunity for grain growth. Whilst the use of microwaves does permit significantly finer grain sizes to be retained (Figure 4.38), a lot more work needs to be done if grain sizes < 100nm are to be achieved via pressureless sintering techniques.

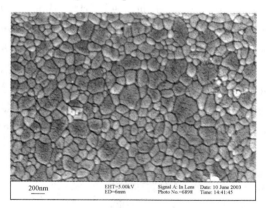

Figure 4.38 Nano Y-PSZ microwave sintered to above 97% of T. D. after 30 min at 1250℃

There is a wide range of microwave absorption characteristics amongst ceramic materials. Most electroceramics, e. g. titanates, and magnetic ceramics, e. g. ferrites, couple extremely well, though the latter interact with the magnetic component of the electromagnetic field. Some engineering ceramics, e. g. zirconia and silicon carbide, also couple well whilst low purity aluminas will tend to heat once they reach 700 ~ 900℃ depending on the impurity type and level. Other ceramics such as high purity alumina and silica are almost transparent to microwaves and will not absorb 2.45 GHz microwaves until well over 1000℃. Despite this, one of the most commonly used ceramics for research into microwave sintering is alumina.

Alumina is the most common ceramic and has been widely used in microwave sintering research by many scientists working in the field. Because of its high refractory nature, it is difficult to sinter it to full densification unless suitable sintering aids or some special processing techniques are adopted. Many people have successfully sintered alumina to high densification with the microwaves. High purity alumina is a low loss material and therefore is not a good microwave absorber at room temperature in a 2.45GHz system. However, typical susceptor materials such as SiC or $MoSi_2$ rod are used to preheat alumina samples. In general, full sintering of the alumina using microwave process has been achieved much faster and at lower temperature than in case of the conventional process. Small disc alumina samples microwave sintered at 1400℃, with no hold time, were 98% dense. Conventional heating requires at 1600℃ and 2h of soaking time to achieve the same degree of densification and substantial grain growth[84].

S. Lefeuvre et al[85] studied the microwave sintering of micro-sized and nano-sized alumina powder in a microwave oven fed by two 1 kW, 2.45 GHz magnetrons in an isolated chamber. The results indicated that it is convenient to mix microparticles and nanoparticles so that the smallest grains fill the space between grains of bigger size. Microwave heating is well-suited to sintering compact of this type because the threshold process which appears in the nanoparticles induces homogeneous heating, further relayed by electrical conductivity in the entire compact.

Recently, Xinhua Ceramic Ltd. (Hunan, China) has successfully used the continuous microwave sintering furnace to produce the alumina electronic ceramic products. They have enjoyed the great benefit from this sintering process. The firing temperature reduced 200℃, the firing time saved 4 h comparing to that by conventional sintering process. It is more important that the characteristics and microstructure of the products have been promoted.

S. M. Vidyavathy and V. Kamaraj[86] reported a result about microwave sintering of Niobium co-doped yttria stabilized zirconia. The combination of transformation toughening and ductile particle reinforcements improves substantially the crack growth resistance and flaw tolerance. The presence of Nb_2O_5 is beneficial to tetragonal to monoclinic transformation. Dense, uniform and fine microstructure was obtained using microwave sintering. Decrease in grain size is attributed to fast heating and cooling rate in the case of microwave sintering.

In the last decades, various researchers have worked with a variety of advanced ceramics such as alumina, zirconia, hydroxyapatite, transparent ceramics, electro-ceramics, ceramic superconductors, glass-ceramics, non-oxide ceramics including Si_3N_4, AlN, SiC etc. In all of these materials, substantial improvements in their properties conventional products were reported. Many related research papers have been published. Hereunder, only a few selected ceramic materials are introduced. All these materials have been processed mainly using a multimode microwaves system operating at 2.45GHz.

Transparent ceramics is an important optical property of many materials. The nature of the material including grain size, density, crystal structure, porosity and the grain boundary phase are the factors that influence the degree of transparency. Most non-cubic ceramics are anisotropic and polycrystalline. The grain boundaries in a ceramic strongly scatter light. Therefore, to convert a non-cubic ceramic having grain larger than the wavelength of light into a transparent ceramic, one must have low grain boundary volume and no intergranular or intragranular porosity. However, if the grain size is smaller than the wavelength of the light (0.4~0.7μm), the light can transmit through the ceramic. Cubic ceramic materials such as spinels and ALON can be made into transparent ceramics even if the grain size is larger than the wavelength of light. To achieve transparency in a ceramic, one must control the grain growth, eliminate porosity and achieve complete densification. The conventional methods to fabricate fully dense and reasonably transparent ceramics involve high temperatures, lengthy sintering conditions, and various complex processing steps, which make the processing of transparent ceramics very difficult and uneconomical. However, the microwave method has been successfully used to fabricate transparent ceramics due to its ability to minimize the grain growth and produce a fully dense ceramic in a very short period of time without utilizing high pressure conditions.

Transparent mullite ceramics have been fabricated in 1996[87]. Fully transparent ALON ceramics were also made using multimode microwave system at 1800℃[88]. Translucent ceramics of AlN, which is a well-known high thermal conductivity material, were also developed in microwave at 1900℃ in 60 min[89].

Recently, we have investigated the microwave sintering and kinetic analysis of densification behavior of Y_2O_3-MgO composites[90]. Compared with conventional sintering, microwave sintering process can effectively promote the densification behavior. At same sintering condition of 1350℃ dwelling 30min, the grain size of sample sintered by conventional process is observably large than that by microwave process, as shown in Figure 4.39. The values of grain growth exponent for microwave sintering indicate that grain boundary diffusion is the main diffusion mechanism during microwave sintering densification process, while volume diffusion plays a main role in the conventional sintering process. We also found that the calculated grain growth activation energy for microwave sintering (108.22kJ/mol) is much lower than that for conventional sintering (160.42kJ/mol). Based on the kinetic analysis data, submicron-grained (about 300nm) Y_2O_3-MgO composites with Vickers hardness of 11.2 ± 0.3GPa were achieved by microwave sintering process, when samples are heated up to 1380℃ with a dwell time of 2 min at the first step, and then cooled to 1285℃ with a dwell time of 55 min at the second step.

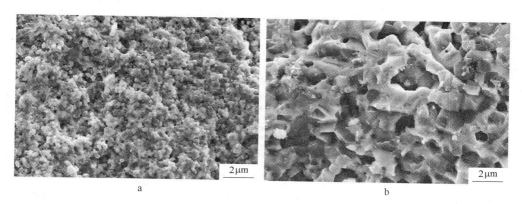

Figure 4.39　SEM micrographs of Y_2O_3-MgO composites sintered by micro-wave sintering (a) and by conventional sintering (b)

$BaTiO_3$ is well recognized that virtually all solid state reactions for the synthesis of materials in a conventional process occur under isothermal conditions, i.e. two or more phases involved are at the same temperature. However, the microwave processing for materials synthesis involving two or more phases may experience a situation known as "anisothermal" state if the reacting phases have different microwave absorption characteristics. The anisothermal situation is associated with huge temperature differences between the phases at micro level. This is also one of the key factors to experience dramatic enhancements in reaction and material diffusion rates. In synthesis of PZT and Ba-TiO_3, the anisothermal approach enhanced reactivity between the starting phases and produced the desired phase in a few minutes. For the synthesis of $BaTiO_3$, $BaCO_3$ (a poor microwave absorber) and TiO_2 (an excellent microwave absorber) were used as the precursors. The reaction of the mixture of these two phases in a microwave field is radically different from the conventional isothermal heating situation. In the microwave case, at 250℃ with no soak tome, hexagonal BaTiO$_3$ appears and at 900℃ in 5 min nearly pure tetragonal BaTiO$_3$ phase is formed. On the other hand, the conventional process——even at 1300℃ for 1h soaking time——does not produced any XRD detectable BaTiO$_3$ phase.

Rui BAO et al[91] processed WC-8Co cemented carbide samples via microwave irradiation in a 2.45GHz, high-power multi-mode microwave cavity. Their results demonstrate that the liquid phase is formed around 1300℃ and nearly full densification is obtained at 1450℃ for 5min. The microstructures of microwave sintered samples have finer and more uniform WC grains than those of vacuum sintered samples.

K. Rajeswari et al[92] reported a comparative evaluation of spark plasma, microwave, two stage sintering❶ and conventional sintering on the densification and micro structural evaluation of fully stabilized zirconia ceramics and drawn following conclusions: Yttria stabilized zirconia specimens sintered through spark plasma sintering at 1325℃ for a period of 5 minutes have exhibited the highest sintered density of 99.89% with lowest grain size of about 1μm. Slip cast specimens sintered by microwave technique at the temperature of 1525℃ have shown a density of 99.15% at an average grain size of 3.7μm with grain size distribution similar to conventional sintered specimens. A significant decrease in grain size to 2.6 μm was observed with the two stage sintering in comparison to conventional sintering and is attributed to the lower soaking temperatures compared to conventional sintering. The microstructure was found to be more homogeneous with a narrow grain size distribution during two stage sintering. Here, the editors of this book regard as that the results of spark plasma sintering and microwave sintering used by K. Rajeswari et al are not comparable, because SPS used powder of fully stabilized zirconia, and MWS used slip casting green body. However, the result of two stage sintering is very interesting.

Now, the principle challenge in the microwave sintering is the ability to measure temperature accurately within a microwave environment. Until this can be achieved then control of temperature, and hence microstructure and properties, cannot be fully optimised. The two principal reasons for the difficulties are: (1) the effect of microwaves

---

❶ Two stage sintering is a sintering method, its schedule is characterized by two regimes wherein the first regime at peak temperature dominates densification and complete elimination of residual porosity followed by a second regime at significantly lower temperatures effecting controlled grain growth during final stages of sintering. Two stage sintering methodology based on the principle that the activation energy for grain growth is lower than the activation energy of densification. The suppression of the final – stage grain growth is achieved by exploiting the difference in kinetics between grain – boundary diffusion and grain – boundary migration.

on thermocouples (they induce currents); (2) the existence of the inverse temperature profile. The latter means that the surface temperature, which most temperature measurement devices record, is lower than the internal temperature which largely controls the densification rate.

Thermocouples can only be used if shielded from the microwaves by a conductive tube. Pyrometry is often used however this again monitors surface temperatures and, as has already been shown, these can be significantly different from internal values. However for research purposes, where measurement of temperature profiles is required, pyrometry is not ideal. An alternative for research purposes is optical thermometry in which the temperature of the sample is sensed by either a microwave transparent black body tip mounted on a sapphire rod or by using a sapphire rod as a light pipe and monitoring temperature by pyrometry.

4.2.9 Compare of Several New Sintering Methods

Nanoceramic sintering has been the driver for the advancement of different sintering techniques. The challenges in densifying nanoceramics while controlling grain growth have pushed the efforts to closely examine the critical sintering factors and design effective sintering processes. Each sintering technique has its own strength as well as limitation to meet the specific needs of nanoceramic sintering. One of the most obvious limitations is the small sample size that nanoceramic sintering can densify, due to the challenge in achieving uniform temperature and pressure, especially when fast heating is involved. Also, nanoceramic sintering needs to improve the processing efficiency with low complexity and cost.

G. Delaizir et al[93] reported a comparative study of spark plasma sintering (SPS), hot isostatic pressing (HIP) and microwaves sintering techniques on p-type Bi_2Te_3 thermoelectric properties. This study demonstrates that the variations of thermoelectric properties between the three samples mainly come from the variations in charge carrier concentration. The highest value of ZT (figure of merit) has been obtained for the ceramic processed by microwaves (ZT = 0.74) and by SPS (ZT = 0.68). The results show that the microwave or SPS techniques are a promising way to sinter thermoelectric materials providing an optimization of the composition of the alloy (amount of antimony) in order to get an optimized charge carrier concentration, this will allow to get higher ZT values.

Shekhar Nath et al[94] reported a comparative study of conventional sintering with microwave sintering of hydroxyapatite synthesized by chemical route. They concluded that for conventional sintering, holding for 2 h at 800°C and subsequent sintering for 3h at

1200℃ result almost fully dense microstructure, characterized by equiaxed grains of 1~2μm size; the microwave sintering, performed by holding at 800℃ for 0.5h and sintering at 1200℃ for 0.5h also produce dense HAp of faceted grains of 1~2μm size; the microwave sintering is found to be a time and energy efficient densification technique in densifying HAp.

Lee Y I et al[95] fabricated dense nanostructured SiC ceramics through two-step sintering, their research results indicated that using a two-step sintering procedure could be prepared nano-structured liquid-phase sintered SiC ceramics with a grain size of about 40 nm. A full densification without notable grain growth was achieved when the specimen was first heated to a higher temperature and then cooled immediately for a long treatment at a lower temperature. The control of density and pore size after the first step is expected to be critical to achieve full densification during the subsequent second step.

T. Thongchai et al[96] (Naresuan University, Pisanuloke, Thailand) reported a comparison of microwave and conventional sintering of Al_2O_3-ZrO_2 composites. The results show that the two different sintering process produce different phase transformation, microwave sintering promoted the formation of t-ZrO_2 at 1300℃ and 1400℃, while this phase forms in sample with conventional sintering at 1400℃. Microwave sintering results in higher densities, flexural and compressive strength compared to conventional sintering. Furthermore by microwave sintering, the microstructure of composite show more dense and uniform grain growth.

P. Figiel et al[97] sintered Al_2O_3 ceramics materials by the conventional free sintering process, 2.45GHz microwave sintering (MW) and spark plasma sintering (SPS), and received alumina ceramic materials with good mechanical and physical properties at the time of 1 minute for microwave sintering to 10 minutes for spark plasma sintering.

Even though spark plasma sintering has many advantages, for example, fast heating, fast cooling and short holding times; to obtain fully dense nanostructures at comparatively low sintering temperatures; to offer very limited grain growth; less sintering additives. However, spark plasma sintering also has its limitations, for example, to be suited to simple shape and small size samples; sintering efficiency depended on the electrical properties of the material being sintered; the expensive equipment; inevitable temperature gradient in the samples and subsequent microstructure inhomogeneity.

Even though microwave sintering can quickly heat the nanoceramics and achieve above 90% density without excessive grain growth, repeated studies have shown that full densification with controlled grain growth by microwave sintering can be a challenging task to achieve. Another problem for microwave sintering is the difficulty of measuring and controlling sintering temperatures.

In order to achieve the maximal densification while avoiding exaggerated grain growth, careful attention must be paid to the sintering time and temperature in the final stage of sintering. Using hybrid heating techniques for the closed pore removal by adapting the pressureless two step sintering strategy maybe is an interesting development direction. This means first microwave sintering the nanostructures to above 90% dense, reducing/removing microwave sintering power, and then using electric resistance heating at low temperatures for extended time in order to achieve full densification while limiting grain growth.

4.3 Control of Sintering

During all the sintering techniques discussed, the densification is controlled by composition, particle size distribution, temperature and time at temperature and pressure (in the case of pressure sintering). Many other factors besides composition, particle size, temperature, and time must be controlled. These factors include atmosphere, time and temperature cycle, setting, and all processing steps of the powder compact prior to sintering. Some of these factors are discussed in the following paragraphs[2].

4.3.1 Firing Curve

A furnace is fired up in three stages: the heating-up stage, the soaking period and the cooling stage.

The variations in the heating-up rate are chosen so that the changes of state to which the product is subject and the stresses that arise from the thermal expansion of the product and the combustion of binders do not cause damage (e.g., cracks and pores).

Furnace design and the heating mechanism determine the maximum heating rate of a furnace. It is important to always check the operating manual that came with a furnace before setting the controller.

The degree of sintering is a function of time and temperature. Different processes use different soaking periods. Table 4.8 shows the corresponding data of firing time vs temperature during firing of 90 alumina ceramics in tunnel kiln.

Table 4.8 Firing procedure of 90 alumina ceramics

t/h	1.5	3	7.5	11.25	12.75	14.25	16.5	18.75	20.25
$T/℃$	19	420	748	1074	1244	1294	1271	1403	1483
t/h	21.75	23.25	24.75	26.25	30	34.5	37.5	39.75	42
$T/℃$	1486	1473	1404	1400	1145	430	379	272	180

Crystallization and other phase transformations that may occur in a product as well as thermal contraction must be taken into account in the cooling stage. At the beginning of the cooling process the material is generally still rather plastic. It can be cooled fairly quickly at this stage, because the thermal contraction does not cause the stresses to increase very much. Below a certain temperature the plastic characteristics disappear. For many clay-based ceramics this is between 800℃ and 600℃. In this range, a higher cooling rate accompanied by a steep temperature gradient can cause stresses to form in the product. In temperature ranges in which phase transformations accompanied by volume changes take place, too high a cooling rate can produce cracks. Especially when using raw materials rich in quartz, the specification of the cooling rate requires special care. One of the goals in processing both traditional and advanced ceramics is to decrease the product throughput time. Fast firing seeks to reduce total costs because of lower energy consumption and improved efficiency.

Time and temperature cycle refers to the rate of heating, peak temperature, the time at peak temperature, and the rate of cooling. Time at temperature and peak temperature influence the total densification achieved and the degree of grain growth. Cooling rate affects the amounts of residual glass in the microstructure. In the case of Mg-PSZ and Ca-PSZ (magnesia and calcia partially stabilized zirconia), the cooling rate and time at selected aging temperatures determine the nature of the strength-controlling and toughness-controlling tetragonal or monoclinic ZrO_2 precipitates. Finally, the heating and cooling rates can affect the mechanical properties of the part being sintered. Too-rapid heat-up or cool-down can cause cracks due to thermal shock. Too-rapid-heating can also cause cracks due to too-rapid burn-off of binders or other organic additives.

Time and temperature cycle and atmosphere must be considered simultaneously for many materials. An example is yttrium barium coprate superconductor. The material must have the fully oxygenated composition $YBa_2Cu_3O_7$ to have optimum superconductivity. $YBa_2Cu_3O_7$, requires a sintering temperature above about 900℃ to achieve full densification. However, an oxygen-deficient tetragonal form is stable above 710℃. An oxygen-deficient structure is retained during normal cooling. Lengthy heat treatment below 710℃ is necessary to achieve a superconductive stoichiometry near $YBa_2Cu_3O_7$. For example, one study determined that a 48h heat treatment at 600℃ was necessary to fully oxygenate a 1mm thick slice of material.

4.3.2 Atmosphere

The atmosphere (gas composition) in the furnace can have obvious or subtle effects on sintering. Obvious effects involve oxidation-reduction conditions. Trying to sinter Si_3N_4, SiC, or other nonoxide compositions in air will result in oxidation. Conversely, trying to sinter oxide ceramics in a carbonaceous or a reducing atmosphere can lead to reduction. For example, ZrO_2 fired in a reducing atmosphere will be oxygen-deficient (ZrO_{2-x}) and will be dark gray or black in color. Residual binder can cause localized reducing conditions and cause sintering to be retarded or accelerated.

Atmosphere can also have subtle effects. The effect on dopants is an example. Many dopants such as Fe, Ce, and Mn can exist in more than one valence (Fe^{3+}, Fe^{2+}, etc). The atmosphere during sintering must be carefully controlled to achieve the desired valence and desired properties.

4.3.3 Sintering Problems

A variety of conditions can result in improper sintering and have a deleterious effect on the material properties. Normally, the manufacturer will detect these problems either during processing or during quality-control inspection. However, sometimes defective or inferior material is not detected by the manufacturer and is shipped to the user, where the defect does not show up until the component fails prematurely in service. Under these circumstances the source of the problem and a feasible solution must be found quickly. The responsible engineer will have a distinct advantage if he or she knows generally how the ceramic was processed and knows what possible problems to look for. The following paragraphs describe some of the problems that can occur during sintering and some of the artifacts in the ceramic component that will help the engineer to identify the cause.

4.3.3.1 Warpage

Warpage is a common problem and is usually detected before the part is put into service. It increases reject rate and hence the cost per part. It also can cause delays if it arises intermittently. Warpage usually results from inadequate support during sintering or from density variations in the green body. The former can be corrected by shifting the orientation of the part in the furnace or by supporting the part with saggers (refractory, nonreactive ceramic pieces that restrict the component from deforming during sintering). The latter can be corrected only by solving the problem in an earlier processing step that caused the inhomogeneity. The two sources of warpage can usually be distinguished from each other by dimensional inspection or by examination of a polished section of the microstructure. Warpage due to sagging will not show variations in thickness or microstructure across the cross section, but warpage due to density variation will.

4.3.3.2　Overfiring

Overfiring is another of the more common sintering problems with ceramics. It can cause warpage, reaction with surrounding furnace structures, bloating, or excessive grain growth. The first three are usually easy to detect visually. Excessive grain growth is more difficult to detect during routine inspection and may require preparation of a polished surface, etching to accentuate the grain structure, and examination by reflected light microscopy.

However, the presence of large grains is readily visible on a fracture surface at low magnification and can provide the engineer with valuable insight into the cause of a component failure. An increase in grain size usually results in a decrease in strength. This is true even if only a portion of the grains have increased size. Sometimes over firing results in exaggerated grain growth, whereby a few grains preferentially grow very large compared to other grains in the microstructure and compared to the optimum grain size required for the intended application.

4.3.3.3　Burn-off of binders

Binders are often added to the ceramic powder prior to compaction. These are usually organic and can leave a carbon residue in the ceramic during sintering if the time and temperature and atmosphere parameters are not properly controlled. If large percentages of binders are present, such as in injection-molded ceramics, the binder may have to be removed very slowly as a gas or liquid by thermal decomposition or capillary extraction. Too-rapid removal results in formation of cracks in the component.

Proper binder removal is normally accomplished by slowly raising the temperature to a level at which the binder can volatilize, and holding at this temperature until the binder is gone. The temperature can then be safely increased to the sintering temperature. However, if the temperature is increased before the binder has completely volatilized, the portion remaining will char and leave a residue of carbon.

In some materials, the carbon will be relatively inert, but in others, it can cause severe chemical reactions during sintering. In one case the carbon resulted in localized reducing conditions in the core of a part, causing bloating and severe dark discoloration. The surface of the same part was white and sintered properly.

4.3.3.4　Decomposition reactions

Ceramics are frequently prepared using a different starting composition than the final composition. For instance, carbonates, sulfates, nitrates, or other salts are often used rather than the oxides, even though the final product is an oxide. There are a variety of reasons for doing this. The salts are often purer or more reactive or can be mixed more uniformly. However, during sintering the salt must decompose to the oxide and react with other constituents to form the desired final composition. If the salt does not decompose early enough, the component can be damaged by gas evolution. If the salt does not decompose completely, an off-composition or inhomogeneous condition can result. The degree of sensitivity of a component to this is dependent on the sintering temperature, the time temperature schedule, and the decomposition temperature and kinetics of the salt. Problems are usually not encountered with hydrates and nitrates because they have low decomposition temperatures. Carbonates tend to have higher decomposition temperatures, but usually do not pose a problem if the sintering temperatures is above 1000℃. Some sulfates do not completely decompose until 1200 to 1300℃.

In summary, sintering is a complex processing, but it is a best important process in the production of ceramic product. All the sintering parameters including sintering temperature, time, dwelling, atmosphere, pressure and additives, and loading pattern must be precisely controlled.

Study Guide

4-1 What is the meaning of the term sintering?
4-2 What are the primary mechanisms of sintering?
4-3 Describe the physical changes that occur during each stage of sintering.
4-4 What are some of the key factors that influence the rate of solid state sintering?
4-5 Explain the factors that lead to grain growth.
4-6 Explain the driving force for solid state sintering.
4-7 Explain how the presence of a liquid phase influences sintering.
4-8 What is reaction sintering?
4-9 What are some problems or defects that can result during sintering?
4-10 What are some advantages and disadvantages of hot pressing?
4-11 Describe how silicon nitride and silicon carbide materials can be fabricated by reaction sintering processes.
4-12 What are some advantages and disadvantages of microwave sintering?
4-13 Explain the reasons of using pressureless sintering.
4-14 Compare the advantages and disadvantages of traditional sintering, spark plasma sintering (SPS) and microwaves sintering techniques.

References

[1] Carter Barry C, Norton Grant M. Ceramic Materials, Science and Engineering [M]. Springer

Science + Business media, LLC. , 2007.
[2] David W Richerson. Modern Ceramic Engineering [M]. Third Edition. Taylor & Francis Group, 2006.
[3] Suk-Joong L Kang. Sintering, Densification, Grain Growth, and Microstructure [M]. Elsevier Butterworth-Heinemann, 2005.
[4] Arthur Dodd. Dictionary of Ceramics [M]. Third Edition. The Institute of Materials, 1994.
[5] Menezes R R, Souto P M, Kiminami R. Microwave Fast Sintering of Ceramic Materials [J]. Sintering of Ceramics——New Emerging Techniques, In Tech, 2012: 3 ~26.
[6] Das S, Mukhopadhyay A K, Datta S, et al. Prospects of Microwave Processing: An Overview [J]. Bulletin of Materials Science, 2009, 32 (1): 1 ~13.
[7] Thridandapani R R. The Effect of Microwave Energy on Sintering [D]. Virginia Polytechnic Institute and State University, 2010.
[8] Zhang Y J, Tan S L, Tan X Y, et al. A Aluminium Silicate Fiber/Potassium Hexatitanate Whiskers Thermal Insulator and Preparation Method: China ZL200510104227. 1 [P]. 2008 - 5 - 28.
[9] Zhang Y J, Gong H Y, Tan S L, et al. A Thermal Insulator with Low Thermal Conductivity and Preparation Method: China, ZL200810138024. 8 [P]. 2009 - 9 - 23.
[10] Riedel R, Ionescu E, Chen I W. Modern Trends in Advanced Ceramics [J]. Ceramics Science and Technology, 2011, 1: 3 ~38.
[11] Rahaman M N. Ceramic Processing and Sintering [M]. Marcel Dekken Inc. , 2003.
[12] Biswas K, Rixecker G, Wiedmann I, et al. Liquid Phase Sintering and Microstructure-property Relationships of Silicon Carbide Ceramics with Oxynitride Additives [J]. Materials chemistry and physics, 2001, 67 (1): 180 ~191.
[13] Svante Prochazka. Hot Pressed Silicon Carbide: U. S. , 3, 853, 566 [P]. 1974 - 10 - 10.
[14] Stutz D H, PROCHAZKA S, LORENZ J. Sintering and Microstructure Formation of β-Silicon Carbide [J]. Journal of the American Ceramic Society, 1985, 68 (9): 479 ~482.
[15] Alliegro R A, Coffin L B, Tinklepaugh J R. Pressure-Sintered Silicon Carbide [J]. Journal of the American Ceramic Society, 1956, 39 (11): 386 ~389.
[16] Magnani G, Betrami G, Minoccari G L, et al. Pressureless Sintering and Properties of α-SiC-B_4C Composite [J]. Journal of the European Ceramic Society, 2001 (21): 633 ~638.
[17] Vassen R, Stöver D. Processing and Properties of Nanograin Silicon Carbide [J]. Journal of the American Ceramic Society, 1999, 82 (10): 2585 ~2593.
[18] Silicon Carbide Ceramics: Fundamental and Solid Reaction [M]. Springer, 1991: 194.
[19] Strecker K, Ribeiro S, Camargo D, et al. Liquid Phase Sintering of Silicon Carbide with AlN/Y_2O_3, Al_2O_3/Y_2O_3 and SiO_2/Y_2O_3 Additions [J]. Materials Research, 1999, 2 (4): 249 ~254.
[20] Magnani G, Beaulardi L, Piiotti L. Properties of Liquid Phase Pressureless Sintered Silicon Carbide Obtained without Sintering Bed [J]. Journal of the European Ceramic Society, 2005,

25 (9): 1619~1627.

[21] Gomezl E, Echeberria J, Iturriza I, et al. Liquid Phase Sintering of SiC with Additions of Y_2O_3, Al_2O_3 and SiO_2 [J]. Journal of the European Ceramic Society, 2004 (24): 2895~2903.

[22] Nagano T, Kaneko K, Zhan G D, et al. Superplasticity of Liquid-Phase-Sintered β-SiC with Al_2O_3-Y_2O_3-AlN Additions in an N_2 Atmosphere [J]. Journal of the American Ceramic Society, 2000, 83 (10): 2497~2502.

[23] Guo X, Yang H, Zhang L, et al. Sintering Behavior, Microstructure and Mechanical Properties of Silicon Carbide Ceramics Containing Different Nano-TiN Additive [J]. Ceramics International, 2010, 36 (1): 161~165.

[24] Rodríguez-Rojas F, Ortiz A L, Guiberteau F, et al. Oxidation Behaviour of Pressureless Liquid-phase-sintered α-SiC with Additions of $5Al_2O_3 + 3RE_2O_3$ (RE = La, Nd, Y, Er, Tm, or Yb) [J]. Journal of the European Ceramic Society, 2010, 30 (15): 3209~3217.

[25] Biswas K. Liquid Phase Sintering of SiC Ceramics with Rare Earth Sesquioxides [D]. Stuttgart University, 2002.

[26] Schwetz K A, Vogt G. Process for the Production of Dense Sintered Shaped Articles of Polycrystalline Boron Carbide by Pressureless Sintering: U. S., 4, 195, 066 [P]. 1980-3-25.

[27] Suzuki H, Hase T, Maruyama T. Effect of Carbon on Sintering of Boron Carbide [J]. J. Ceram. Soc. Jpn., 1979, 87 (1008): 430~433.

[28] Schwetz K A, Grellner W. The Influence of Carbon on the Microstructure and Mechanical Properties of Sintered Boron Carbide [J]. Journal of the Less Common Metals, 1981, 82: 37~47.

[29] Lee H, Speyer R F. Pressureless Sintering of Boron Carbide [J]. Journal of the American Ceramic Society, 2003, 86 (9): 1468~1473.

[30] Bougoin M, Thevenot F. Pressureless Sintering of Boron Carbide with an Addition of polycarbosilane [J]. Journal of Materials Science, 1987, 22 (1): 109~114.

[31] Mashhadi M, Taheri-Nassaj E, Sglavo V M, et al. Effect of Al Addition on Pressureless Sintering of B_4C [J]. Ceramics International, 2009, 35 (2): 831~837.

[32] Kumazawa T, Honda T, Zhou Y, et al. Pressureless Sintering of Boron Carbide Ceramics [J]. Journal of the Ceramic Society of Japan, 2008, 116 (1360): 1319~1321.

[33] Frage N, Hayun S, Kalabukhov S, et al. The Effect of Fe Addition on the Densification of B_4C Powder by Spark Plasma Sintering [J]. Powder Metallurgy and Metal Ceramics, 2007, 46 (11~12): 533~538.

[34] Goldstein A, Yeshurun Y, Goldenberg A. B_4C/Metal Boride Composites Derived from B_4C/Metal Oxide Mixtures [J]. Journal of the European Ceramic Society, 2007, 27 (2): 695~700.

[35] Yamada S, Hirao K, Yamauchi Y, et al. Densification Behaviour and Mechanical Properties of Pressureless-sintered B_4C-CrB_2 Ceramics [J]. Journal of Materials Science, 2002, 37 (23): 5007~5012.

[36] Baharvandi H R, Hadian A M, Alizadeh A. Processing and Mechanical Properties of Boron Carbide-titanium Diboride Ceramic Matrix Composites [J]. Applied Composite Materials, 2006, 13 (3): 191~198.

[37] Wei R B, Zhang Y J, Gong H Y, et al. The Effect of Rare-earth Oxide Addition on the Densification of Pressureless Sintering B_4C Ceramics [J]. Ceramics International, 2013 (39): 6449~6452.

[38] Fritz Aldinger, Nils Claussen, Richard M Spriggs, et al. Handbook of Advanced Ceramics [M]. Elsevier Inc., 2003.

[39] Bocanegra-Bernal M H, Matovic B. Dense and Near-net-shape Fabrication of Si_3N_4 Ceramics [J]. Materials Science and Engineering A, 2009 (500): 130~149.

[40] Zhao D L, Zhang Y J, Gong H Y, et al. Effects of Sintering Aids on Mechanical and Dielectric Properties of Si_3N_4 Ceramics [J]. Materials Research Innovations, 2010, 14 (4): 338~341.

[41] Mishra R S, Risbud S H, Mukherjee A K. Influence of Initial Crystal Structure and Electrical Pulsing on Densification of Nanocrystalline Alumina Powder [J]. Journal of Materials Research, 1998, 13 (01): 86~89.

[42] Zhou Y, Hirao K, Toriyama M, et al. Very Rapid Densification of Nanometer Silicon Carbide Powder by Pulse Electric Current Sintering [J]. Journal of the American Ceramic Society, 2000, 83 (3): 654~656.

[43] Mishra R S, Mukherjee A K. Electric Pulse Assisted Rapid Consolidation of Ultrafine Grained Alumina Matrix Composites [J]. Materials Science and Engineering: A, 2000, 287 (2): 178~182.

[44] Hong P. Spark Plasma Sintering of Si_3N_4-Based Ceramics-Sintering Mechanism-Tailoring Microstructure-Evaluating Properties [D]. Stockholm University, 2004.

[45] Tokita M. Mechanism of Spark Plasma Sintering [C] //Proceeding of NEDO International Symposium on Functionally Graded Materials. Japan, 1999, 21: 22.

[46] Hungria T, Galy J, Castro A. Spark Plasma Sintering as a Useful Technique to the Nanostructuration of Piezo-Ferroelectric Materials [J]. Advanced Engineering Materials, 2009, 11 (8): 615~631.

[47] Maitre A, Put A V, Laval J P, et al. Role of Boron on the Spark Plasma Sintering of an α-SiC powder [J]. Journal of the European Ceramic Society, 2008, 28 (9): 1881~1890.

[48] Xu C, Cai Y, Flodström K, et al. Spark Plasma Sintering of B_4C Ceramics: The Effects of Milling Medium and TiB_2 Addition [J]. International Journal of Refractory Metals and Hard Materials, 2012, 30 (1): 139~144.

[49] Hayun S, Paris V, Dariel M P, et al. Static and Dynamic Mechanical Properties of Boron Carbide Processed by Spark Plasma Sintering [J]. Journal of the European Ceramic Society, 2009, 29 (16): 3395~3400.

[50] Kima K H, Chaea J H, Parka J S, et al. Sintering Behavior and Mechanical Properties of B_4C

Ceramics Fabricated by Spark Plasma Sintering [J]. Journal of Ceramic Processing Research, 2009, 10 (6): 716~720.

[51] Frage N, Hayun S, Kalabukhov S, et al. The Effect of Fe Addition on the Densification of B_4C Powder by Spark Plasma Sintering [J]. Powder Metallurgy and Metal Ceramics, 2007, 46 (11~12): 533~538.

[52] Ryu H J, Lee Y W, Cha S I, et al. Sintering Behaviour and Microstructures of Carbides and Nitrides for the Inert Matrix Fuel by Spark Plasma Sintering [J]. Journal of Nuclear Materials, 2006, 352 (1): 341~348.

[53] Xiong Y, Fu Z Y, Wang H, et al. Microstructure and IR Transmittance of Spark Plasma Sintering Translucent AlN Ceramics with CaF_2 Additive [J]. Materials Science and Engineering: B, 2005, 123 (1): 57~62.

[54] He X, Ye F, Zhang H, et al. Study of Rare-earth Oxide Sintering Additive Systems for Spark Plasma Sintering AlN Ceramics [J]. Materials Science and Engineering: A, 2010, 527 (20): 5268~5272.

[55] Lee Y I, Lee J H, Hong S H, et al. Preparation of Nanostructured TiO_2 Ceramics by Spark Plasma Sintering [J]. Materials Research Bulletin, 2003, 38 (6): 925~930.

[56] Kim B N, Hiraga K, Morita K, et al. Effects of Heating Rate on Microstructure and Transparency of Spark-plasma-sintered Alumina [J]. Journal of the European Ceramic Society, 2009, 29 (2): 323~327.

[57] Pravarthana D, Chateigner D, Lutterotti L, et al. Growth and Texture of Spark Plasma Sintered Al_2O_3 Ceramics: A Combined Analysis of X-rays and Electron back Scatter Diffraction [J]. Journal of Applied Physics, 2013, 113 (15): 153510.

[58] Jiang D T, Hulbert D M, Anselmi-Tamburini U, et al. Optically Transparent Polycrystalline Al_2O_3 Produced by Spark Plasma Sintering [J]. Journal of the American Ceramic Society, 2008, 91 (1): 151~154.

[59] Chaim R, Shen Z, Nygren M. Transparent Nanocrystalline MgO by Rapid and Low-temperature Spark Plasma Sintering [J]. Journal of Materials Research, 2004, 19 (9): 2527~2531.

[60] Chaim R, Estournes C. Densification of Nanocrystalline Y_2O_3 Ceramic Powder by Spark Plasma Sintering [J]. Journal of the European Ceramic Society, 2009 (29): 91~98.

[61] Meir S. Fabrication of Transparent Magnesium Aluminate Spinel by the Spark Plasma Sintering Technique [D]. Ben-Gurion University of the Negev, 2008.

[62] Zamora V, Ortiz A L, Guiberteau F, et al. Spark-plasma Sintering of ZrB_2 Ultra-high-temperature Ceramics at Lower Temperature via Nanoscale Crystal Refinement [J]. Journal of the European Ceramic Society, 2012, 32 (10): 2529~2536.

[63] Abe J, Kobune M, Kitada K, et al. Effects of Spark-Plasma Sintering on the Piezoelectric Properties of High-Density $(1-x)(Na_{0.5}K_{0.5})NbO_{3-x}LiTaO_3$ Ceramics [J]. Journal of Korean Physical Society, 2007, 51 (2): 810~814.

[64] Lee B T, Kim K H, Rahman A H M E, et al. Microstructures and Mechanical Properties of

Spark Plasma Sintered Al_2O_3-Co Composites Using Electroless Deposited Al_2O_3-Co Powders [J]. Materials Transactions, 2008, 49 (6): 1451~1455.

[65] Mali V I, Anisimov A G, Kurguzov V D, et al. Spark Plasma Sintering for the production of micron-and nanoscale materials. http://www.researchgate.net/publication/260105435.

[66] Taslicukur Z, Sahin F C, Kuskonmaz N. Properties of Si_3N_4/SiC Composites Produced via Spark Plasma Sintering [J]. International Journal of Materials Research, 2012, 103 (11): 1337~1339.

[67] Peng Y, Peng Z, Ren X, et al. Effect of SiC Nano-whisker Addition on TiCN-based Cermets Prepared by Spark Plasma Sintering [J]. International Journal of Refractory Metals and Hard Materials, 2012, 34: 36~40

[68] Cheng B, Lin Y, Lan J, et al. Preparation of In_2O_3-Sr_2RuErO_6 Composite Ceramics by the Spark Plasma Sintering and Their Thermoelectric Performance [J]. Journal of Materials Science & Technology, 2011, 27 (12): 1165~1168.

[69] Zhan G D, Kuntz J D, Duan R G, et al. Spark-Plasma Sintering of Silicon Carbide Whiskers (SiCw) Reinforced Nanocrystalline Alumina [J]. Journal of the American Ceramic Society, 2004, 87 (12): 2297~2300.

[70] Hulbert D M, Jiang D, Dudina D V, et al. The Synthesis and Consolidation of Hard Materials by Spark Plasma Sintering [J]. International Journal of Refractory Metals and Hard Materials, 2009, 27 (2): 367~375.

[71] Kljajević L, Nenadović S, Nenadović M, et al. Spark Plasma Sintering of ZrC-SiC Ceramics with $LiYO_2$ Additive [J]. Ceramics International, 2013, 39 (5): 5467~5476.

[72] Miranzo P, Ramírez C, Román-Manso B, et al. In Situ Processing of Electrically Conducting Graphene/SiC Nanocomposites [J]. Journal of the European Ceramic Society, 2013, 33 (10): 1665~1674.

[73] Wang L, Wu T, Jiang W, et al. Novel Fabrication Route to Al_2O_3-TiN Nanocomposites via Spark Plasma Sintering [J]. Journal of the American Ceramic Society, 2006, 89 (5): 1540~1543.

[74] Wang L, Jiang W, Chen L, et al. Rapid Reactive Synthesis and Sintering of Submicron TiC/SiC Composites through Spark Plasma Sintering [J]. Journal of the American Ceramic Society, 2004, 87 (6): 1157~1160.

[75] Zhang J, Wang L, Shi L, et al. Rapid Fabrication of Ti_3SiC_2-SiC Nanocomposite Using the Spark Plasma Sintering-reactive Synthesis (SPS-RS) method [J]. Scripta Materialia, 2007, 56 (3): 241~244.

[76] Anselmi-Tamburini U, Kodera Y, Gasch M, et al. Synthesis and Characterization of Dense Ultra-high Temperature Thermal Protection Materials Produced by Field Activation through Spark Plasma Sintering (SPS): I. Hafnium Diboride [J]. Journal of Materials Science, 2006, 41 (10): 3097~3104.

[77] Nygren M. SPS Processing of Nano-structured Ceramics [J]. Journal of Iron and Steel Re-

search, International, 2007, 14 (5): 99~103.
[78] Lu K. Sintering of Nanoceramics [J]. International Materials Reviews, 2008, 53 (1): 21~38.
[79] Gephart S. Field Assisted Sintering of Silicon Carbide: Effects of Temperature, Pressure, Heating Rate, and Holding Time [D]. The Pennsylvania State University, 2010.
[80] Arunachalam Lakshmanan. Sintering of Ceramic-New Emerging Techniques [M]. Publisher In Tech, 2012.
[81] Binner J, Vaidhyanathan B. Microwave Sintering of Ceramics: What does It Offer? [J]. Key Engineering Materials, 2004, 264: 725~730.
[82] Sujith A V, Amar Kumar N, Sharan N. Microwave Sintering of Zirconia and Alumina [J]. International Journal of Recent Trends in Engineering, 2009, 1 (3): 320~323.
[83] Janney M A, Calhoun C L, Kimrey H D. Microwave Sintering of Solid Oxide Fuel Cell Materials: I, Zirconia-8mol% Yttria [J]. Journal of the American Ceramic Society, 1992, 75 (2): 341~346.
[84] Agrawal D. Microwave Sintering of Ceramics, Composites and Metallic Materials, and Melting of Glasses [J]. Transactions of the Indian Ceramic Society, 2006, 65 (3): 129~144.
[85] Lefeuvre S, Federova E, Gomonova O, et al. Microwave Sintering of Micro-and Nano-Sized Alumina Powder [J]. Advances in Modeling of Microwave sintering, 2010: 8~9.
[86] Vidyavathy S M, Kamaraj V. Microwave Sintering of Niobium Co-doped Yttria Stabilized Zirconia [J]. Modern Applied Science, 2009, 3 (6): 102~104.
[87] Fang Y, Roy R, Agrawal D K, et al. Transparent Mullite Ceramics from Diphasic Aerogels by Microwave and Conventional Processings [J]. Materials Letters, 1996, 28 (1): 11~15.
[88] Cheng J, Agrawal D, Zhang Y, et al. Microwave Reactive Sintering to Fully Transparent Aluminum Oxynitride (ALON) Ceramics [J]. Journal of Materials Science Letters, 2001, 20 (1): 77~79.
[89] Cheng J, Agrawal D, Zhang Y, et al. Microwave Sintering of Transparent Alumina [J]. Materials Letters, 2002, 56 (4): 587~592.
[90] Sun H B, Zhang Y J, Gong H Y, et al. Microwave Sintering and Kinetic Analysis of Y_2O_3-MgO Composites [J]. Ceramics International, 2014 (40): 10211~10215.
[91] Bao R, Yi J, Peng Y, et al. Effects of Microwave Sintering Temperature and Soaking Time on Microstructure of WC-8Co [J]. Transactions of Nonferrous Metals Society of China, 2013, 23 (2): 372~376.
[92] Rajeswari K, Hareesh U S, Subasri R, et al. Comparative Evaluation of Spark Plasma (SPS), Microwave (MWS), Two Stage Sintering (TSS) and Conventional Sintering (CRH) on the Densification and Micro Structural Evolution of Fully Stabilized Zirconia Ceramics [J]. Science of Sintering, 2010, 42 (3): 259~267.
[93] Delaizir G, Bernard-Granger G, Monnier J, et al. A Comparative Study of Spark Plasma Sintering (SPS), Hot Isostatic Pressing (HIP) and Microwaves Sintering Techniques on P-type

Bi_2Te_3 Thermoelectric Properties [J]. Materials Research Bulletin, 2012, 47 (8): 1954~1960.

[94] Nath S, Basu B, Sinha A. A Comparative Study of Conventional Sintering with Microwave Sintering of Hydroxyapatite Synthesized by Chemical Route [J]. Trends in Biomaterials & Artificial Organs, 2006, 19 (2): 93~98.

[95] Lee Y I, Kim Y W, Mitomo M, et al. Fabrication of Dense Nanostructured Silicon Carbide Ceramics through Two-step Sintering [J]. Journal of the American Ceramic Society, 2003, 86 (10): 1803~1805.

[96] Thongchai T, Larpkiattaworn S, Atong D, et al, Comparison of Microwave and conventional sintering of Al_2O_3-ZrO_2 composites. [18] The international conference on composite materials.

[97] Figiel P, Rozmus M, Smuk B. Properties of Alumina Ceramics Obtained by Conventional and Non-conventional Methods for Sintering Ceramics [J]. Journal of Achievements in Materials and Manufacturing Engineering, 2011, 48 (1): 29~34.